MIX
Papier aus verantwortungsvollen Quellen
Paper from responsible sources
FSC® C105338

Pritesh Ranjan Dash
Kanzil Mowla Mou

A Comprehensive Review on Five Medicinal Plants of Bangladesh

Chemical Constituents and Uses

Anchor Academic
Publishing

Dash, Pritesh Ranjan, Mou, Kanzil Mowla: A Comprehensive Review on Five Medicinal Plants
of Bangladesh. Chemical Constituents and Uses, Hamburg, Anchor Academic Publishing
2017

Buch-ISBN: 978-3-96067-117-6
PDF-eBook-ISBN: 978-3-96067-617-1
Druck/Herstellung: Anchor Academic Publishing, Hamburg, 2017

Bibliografische Information der Deutschen Nationalbibliothek:
Die Deutsche Nationalbibliothek verzeichnet diese Publikation in der Deutschen
Nationalbibliografie; detaillierte bibliografische Daten sind im Internet über
http://dnb.d-nb.de abrufbar.

Bibliographical Information of the German National Library:
The German National Library lists this publication in the German National Bibliography.
Detailed bibliographic data can be found at: http://dnb.d-nb.de

All rights reserved. This publication may not be reproduced, stored in a retrieval system or
transmitted, in any form or by any means, electronic, mechanical, photocopying, recording or
otherwise, without the prior permission of the publishers.

Das Werk einschließlich aller seiner Teile ist urheberrechtlich geschützt. Jede Verwertung
außerhalb der Grenzen des Urheberrechtsgesetzes ist ohne Zustimmung des Verlages
unzulässig und strafbar. Dies gilt insbesondere für Vervielfältigungen, Übersetzungen,
Mikroverfilmungen und die Einspeicherung und Bearbeitung in elektronischen Systemen.

Die Wiedergabe von Gebrauchsnamen, Handelsnamen, Warenbezeichnungen usw. in diesem
Werk berechtigt auch ohne besondere Kennzeichnung nicht zu der Annahme, dass solche
Namen im Sinne der Warenzeichen- und Markenschutz-Gesetzgebung als frei zu betrachten
wären und daher von jedermann benutzt werden dürften.

Die Informationen in diesem Werk wurden mit Sorgfalt erarbeitet. Dennoch können Fehler nicht
vollständig ausgeschlossen werden und die Diplomica Verlag GmbH, die Autoren oder
Übersetzer übernehmen keine juristische Verantwortung oder irgendeine Haftung für evtl.
verbliebene fehlerhafte Angaben und deren Folgen.

Alle Rechte vorbehalten

© Anchor Academic Publishing, Imprint der Diplomica Verlag GmbH
Hermannstal 119k, 22119 Hamburg
http://www.diplomica-verlag.de, Hamburg 2017
Printed in Germany

Abstract:

Since primordial period, plants are being utilized as a potent source of medicine to treat many life-threatening diseases. One of the potential way to evaluate the importance of the medicinal plant is to identify its active chemical constituents and pharmacological activities. Thus, the present study involves a thorough discussion about the general description, phytochemistry and medicinal properties of five different plants. *Gymnema sylvestre*, *Momordica charantia*, *Coccinia cordifolia*, *Trigonella foenum-graecum*, and *Lagerstroemia speciosa* are the selected plants which are the main focus of the study. All the selected plants are belong to different families but possess similar pharmacological activities such as anti-diabetic, anti-cancer, antioxidant, antimicrobial, analgesic, anti-inflammatory, anti-nociceptive, hypolipidemic and so on. Here, we have reviewed all the reported chemical constituents as well as the pharmacological activities of the plants.

Keywords: *Gymnema sylvestre*, *Momordica charantia*, *Coccinia cordifolia*, *Trigonella foenum-graecum*, *Lagerstroemia speciosa*, Phytoconstituents, Pharmacological activities, Toxicity

Table of Contents

Title	Page no.
Chapter 1	
Introduction	01-02
Scientific Names	03
Common Names	03
Botanical Description of the Medicinal Plants	04-05
Taxonomical Classification	06-07
Chapter 2	
Phytochemical constituents	08-22
Gymnema sylvestre	08-10
Momordica charantia	10-12
Coccidian cordifolia	12-16
Trigonella foenum-graecum	16-19
Lagerstroemia speciosa	19-22
Chapter 3	
Mechanism of action of the plants	23
Chapter 4	
Pharmacological Activities of Five Different Plants	24-47
Pharmacological Activities of *Gymnema sylvestre*	24-27
Pharmacological Properties of *Momordica charantia*	27-34
Pharmacological Investigations on *Coccinia Cordifolia*	35-39
Pharmacological Investigations on *Trigonella Foenum-Graecum. L*	39-44
Pharmacological Investigations on *Lagerstromia speciosa*	44-47
Chapter 5	
Conclusion	48
References	48-62

Chapter 1

Introduction:

Plant derived substances are nowadays one of the potent sources of medicine to cure several complicated diseases because of their versatile applications. It is now believed that nature has given the cure of every disease exactly as the way it should be treated. Both in developed and developing countries progressive demand of traditional medicines are explored by scientific research. As a consequence, the researcher nowadays are more emphasizing on evaluation and characterization of numerous medicinal plants and plant constituents against variety of diseases based on their traditional claims. It is worth mentioning that the present study thus expressing phytochemical constituents and pharmacological activities of five different medicinal plants having potential medicinal properties. In this regards, *Gymnema sylvestre, Momordica charantia, Coccidian cordifolia, Trigonella foenum-graecum,* and *Lagerstroemia speciosa* are such plants. *Gymnema sylvestre*, basically an anti-diabetic plant belonging to the family of Asclepiadaceae; commonly distributed throughout the world, predominantly in tropical regions. Because of upholding diversified medicinal properties, its plant parts can be used in the treatment of diabetes, helminthiasis, dyspepsia, constipation, jaundice, hemorrhoids, cardiopathy, asthma, bronchitis, leucoderma and several inflammatory diseases. In addition, some of the potential pharmacological activities such as antimicrobial, anti-hyphal, anti-hypercholesterolemic, and hepatoprotective activities (Fabio et al., 2013) of the plant was also reported. Recent investigation on this plant is also included the hypoglycemic and chemo-preventive activities of the plant extract (Kumar, 2016). Another medicinal plant of great importance *Momordica charantia,* belonging to the family of *Cucurbitaceae,* also named as bitter melon, balsam pear and bitter gourd. The fruit of this plant is basically used as vegetable and also has the ability to treat small pox infection. Aqueous extract of *Momordica charantia* is a potent anti-diabetic, anti-hyperlipidemic and anti-caricinogenic. It also possesses anti-HIV, anti-helminthic, anti-tumor and wound healing properties (Ahmad et al., 2014). *Coccinia cordifolia,* also belonging to the family of *Cucurbitaceae* was found to be a potent source in order to treat many diseases. The whole plant of *C. cordifolia* possesses diversified pharmacological activities like analgesic, antipyretic, anti-inflammatory, antimicrobial, antiulcer, antidiabetic, antioxidant, hypoglycemic, hepatoprotective, antimalarial, antidyslipidemic, anticancer, antitussive, mutagenic activities. It is also evident that the ethanolic leaf extract of *Coccinia cordifolia* has strong effect against bacterial strains compare to its root (Gautam et al., 2014).

Trigonella foenum -graecum (Linn.) belonging to the family Papilionaceae commonly known as Fenugreek. Its seeds and leaves are used not only as food but also serve several medical benefits like hypocholesterolemic, lactation aid, antibacterial, antifungal, gastric stimulant, for anorexia, antidiabetic agent, galactogogue, hepatoprotective effect and anticancer. Apart from its medicinal benefits, it can also be utilized as a part of several food product developments as food stabilizer, adhesive, and emulsifying agent. More importantly it is used for the development of healthy and nutritious extruded and bakery product (Wani et al., 2016). Lastly, one more important medicinal plant named *Lagerstroemia speciosa,* which is belongs to Lythraceae family and in India commonly known as queen's flower, queen of flowers, crepe myrtle and pride of India whereas in Philippine referred to as "banaba". As remedy the leaves, roots and bark of *L. speciosa* have been used in folk medicine traditionally for several disorders and ailments. The leaves of this plant can be utilized as a diuretic and decongestant, and have been used to treat diabetes mellitus. Not only its leaves serve medicinal benefits but also its roots have been used to treat mouth ulcers. Even its bark is used as a stimulant, febrifuge, and for relief of abdominal pains. This plant serves different purposes in different places such as in Philippines, *L. speciosa* leaves are consumed as herbal tea for lowering blood sugar level, reducing body weight as well while in India, it is used to treat diabetes. Moreover, for garnishing dishes or as an ingredient in salad, soups, desserts and drinks the flowers of this plant is greatly used. Recently, due to having anti-diabetic property, herbal products like Banabamin and Glucosol TM have been developed from *L. speciosa* after conducting preliminary clinical trials (Chan et al., 2014) The following is a compilation and up-to-date review containing the generalized description, phytochemical constituents and pharmacological properties of the selected plants with an urge of future advancement of the medicinal plants to mitigate human diseases.

.

Scientific Names:

Name of Different Medicinal Plants				
Gymnema sylvestre	*Momordica charantia*	*Coccinia cordifolia*	*Trigonella foenum-graecum*	*Lagerstroemia speciosa*
Asclepias geminate, Asclepias geminata, Periploca sylvestris, Gymnema melicida	*Momordica charantia*	*Coccinia indica, Coccinia cordifolia*	*Trigonella foenum-graecum*	*Lagerstroemia speciosa*

Common Names:

Gymnema sylvestre: Gurmar, Merasingi, Meshashringi, Gurmarbooti, Peiploca of the Woods, Rams' horn, Small Indian ipecac, Sugar destroyer, Meshashringi, Madhunashini, Ajaballi, Ajagandini, Bahalchakshu, Karnika, Chakshurabahala, Kshinavartta, Kavali, Kalikardori, Vakundi, Gurmar, Merasingi, Kavali, Kalikardori, Vakundi, Dhuleti, Mardashingi, Podapatri, Adigam, Cherukurinja, Sarkarikolli, Sannager-asehambu, Chakkarakolli, Madhunashini, Mera-Singi.

Momordica charantia: Bitter melon, Balsam pear, Bitter cucumber, Bitter pear, Karalla, Balsam apple, Cerasee, Carilla cundeamor, Papailla, Melao de sao ceatano, Bitter gourd, Sorosi, Karela, Kurela, Kor-kuey, Pava-aki, Salsamino, Sorossies, Pare, Peria, Karla, Margose, Goo-fah, Mara chean.

Coccinia cordifolia: Telakucha, Kuchla (Beng.) ; Ivy gourd (Eng).

Trigonella foenum-graecum: Methi, Fenugreek, Alhova, Bird's Foot, Greek Clover, Greek Hay.

Lagerstroemia speciosa: Queen's flower, Pride of India, Queen's crape myrtle, Banaba, Jarul.

Botanical Description of the Medicinal Plants:

Gymnema sylvestre: *G. sylvestre* is a slow growing, perennial, woody climber, distributed throughout the India, in dry forests up to 600m height. Mainly it is concentrated in the tropical forest of Central and Southern India. Moreover, it is found in Banda, konkan, Western Ghats, Deccan extending to the parts of western and northern India (Kaviarasan et al., 1990). The plant is a large, more or less pubescent, woody climber. The leaves are opposite, usually elliptic or ovate (1.25 – 2.0 inch x 0.5-1.25 inch). Flowers are small, yellow, in axillary and lateral umbel in cymes; Follicles are terete and lanceolate upto 3 inches in length. The Calyx-lobes are long, ovate, obtuse and pubescent. Corolla is pale yellow campanulate, valvate, corona single, with 5 fleshy scales. Scales adnate to throat of corolla tube between lobes; another connective produced into a memberanous tip, pollinia 2, erect, carpels 2, unilocular; locules many ovuled (Madhurima et al., 2009; Potawale, 2008; Gurav, 2007; Zhen, 2001).

Momordica charantia: Bitter melons may be cultivated in tropical areas including Africa, India, South America, and Asia. It is characterize as a perennial vine bearing yellowish green oblong bumpy fruit resembling a cucumber. Once ripened, it releases brown and white color seeds which are embedded in its red pulp. The herbaceous, tendril-bearing vine grows to 5 m. It bears simple, alternate leaves 4–12 cm across, with 3–7 deeply separated lobes. Each plant bears separate yellow male and female flowers.

Coccinia cordifolia: *C. cordifolia* is a dioecious, perennial and herbaceous climber with glabrous stems and tuberous roots This plant possess axillary tendrils long-lived scrambling or climbing vine grows up to 13 m in height and can form a very dense cover over vegetation. It usually covers trees, understory vegetation, fences, power poles, and other human-made structures in residential neighborhoods and agricultural areas. When stems of *C. cordifolia* touch soil, they strike roots readily at the nodes (Chun, 2001). Initially, younger stems are slender, green, and smooth but as they grow they become swollen and semi-succulent in nature. Leaves are alternate and simple. The alternately arranged leaves are borne on stalks 1-3 cm long and coiled tendrils are often produced in their forks. These lobed leaves are somewhat ivy-shaped in nature (3.5-9 cm long and 4-9 cm wide) and usually have tiny teeth spaced along their margins. The tendrils are long, elastic with coil-like springy character that can wrap around the host to the entire length. This species produces separate male and female flowers on separate plants. These white, tubular, flowers are borne singly in the leaf forks on stalks 1-5 cm long. They have five small narrow sepals (6-8 mm long) that are joined together at the base and usually have five spreading petal lobes with pointed tips. In the short tube at the centre of the male flowers are three convoluted stamens, while the centre of the female flowers usually bears three hairy stigmas. The ivy gourd fruit belongs to the berry type: oval and hairless with thick and sticky skin. The raw fruit is green in color resembles a small dark green cucumber with paler stripes. These fleshy fruit (2.5-6 cm long and up to 3.5 cm wide) turn bright scarlet red as they mature and contains several pale, flattened seeds. Two varieties of *C. cordifolia* are recognized; tender fruits are bitter in one variety and not bitter in other, and the latter is used in Asian cooking (Ramachandaran, 1983; Kunkel, 1984 & Manandhar, 2002). Morphologically no difference is evident between them, however; both varieties are invasive and are found to grow close to each other.

***Trigonella foenum-graecum*:** Fenugreek plant is a quick growing annual leguminous herb about 2 feet in height. Leaves are light green in color. Plant stems are long and slender. Fenugreek leaves are tripartite, toothed, grey-green obovate leaves, 20-25 mm long. Fenugreek plant blooms white flowers in the summer. Fenugreek seed pods contain ten to twenty small, flat, yellow-brown, pungent, aromatic seeds. Fenugreek seeds are small and stony; about 1/8 inch long, oblong, rhomboidal, with a deep furrow dividing them into two unequal lobes. Fenugreek seeds have a strong aroma and bitter in taste. Plants mature in about four months. The whole plant is uprooted and allowed to dry. The seeds are threshed out and further dried.

***Lagerstroemia speciosa*:** It is a small to medium-sized tree growing to 20 m tall, with smooth, flaky bark. The leaves are deciduous, oval to elliptic, 8-15 cm long and 3-7 cm broad, with an acute apex. The flowers are produced in erect panicles 20-40 cm long, each flower with six white to purple petals 2-3.5 cm long.

Taxonomical Classification: (Fabio et al., 2013; Gautam et al., 2014; Yadev et al., 2011 & Flora, The gardeners Bible, 2005)

Gymnema sylvestre	
Kingdom	**Plantae**
Subkingdom	Tracheobionta
Superdivision	Spermatophyta
Division	Magnoliophyta
Class	Magnoliopsida
Subclass	Asteridae
Order	Gentianales
Family	Asclepiadaceae
Genus	*Gymnema* R. Br.
Species	*sylvestre*

Momordica charantia
Kingdom: Plantae
Division: Magnoliophyta
Class: Magnoliopsida
Order: Cucurbitales
Family: Cucurbitaceae
Genus: Momordica
Species: M.charantia

Coccinia cordifolia	
Kingdom	**Plantae**
Order	Cucurbitales
Family	Cucurbitaceae
Genus	*Coccinia*
Species	*Coccinia cordifolia*

Trigonella foenum-graecum
Kingdom: Plantae
Division: Magnoliophyta
Class: Magnoliopsida
Order: Fabales
Family: Fabaceae
Genus: *Trigonella*
Species: *T. foenum-graecum*

Lagerstroemia speciosa
Kingdom: Plantae
Order: Myrtales
Family: Lythraceae
Genus: *Lagerstroemia*
Species: L. *speciosa*

a) Leaves of *Gymnema sylvestre*

b) Whole plant of *Momordica charantia*

c) *Coccinia cordifolia*

d) Plant of *Trigonella foenum-graecum*

e) Leaves and flower of *Lagerstromia speciosa*

Chapter 2

Phytochemical Constituents:

Gymnema sylvestre: The leaves of *G. sylvestre* contain triterpene saponins belonging to oleanane and dammarene classes. Oleanane saponins are gymnemic acids and gymnemasaponins, while dammarene saponins are gymnemasides (Khramov, 2008). The leaves also contain resins, albumin, chlorophyll, carbohydrates, tartaric acid, formic acid, butyric acid, anthraquinone derivatives, inositole alkaloids, organic acid (5.5%), parabin, calcium oxalate (7.3%), lignin (4.8%), and cellulose (22%). A new flavonol glycoside namely kaempferol 3-O-beta-D-glucopyranosyl-(1-->4)- alpha-Lrhamnopyranosyl-(1-->6)- beta-D-galactopyranoside has also been found in aerial parts of *G. sylvestre* (Kuzuku, 1989 & Liu, 2004). In addition, three new oleanane type triterpene glycosides and four new triterpenoid saponins, gymnemasins; have been isolated from the leaves of *G. sylvestre* were identified as 3-O-[beta-D-glucopyranosyl(1-->3)-beta-Dglucopyranosyl]-22-O-tiglyol-gymnemanol,3-O-[beta-D-glucopyranosyl(1-->3)-beta-D-lucuro-nopyranosyl]-gymnemanol, 3-O-beta-D-glucuronopyranosyl-22-O-tigloyl-gymnemanol and 3-O-beta-D-glucopyranosyl-gymnemanol respectively. Moreover, gymnestrogenin, a new pentahydroxytriterpene from the leaves of *G. sylvestre* has also been reported (Saho, 1996). Other plant constituents are flavones, anthraquinones, hentri-acontane, pentatriacontane, α and β-chlorophylls, phytin, resins, d-quercitol, tartaric acid, formic acid, butyric acid, lupeol, β-amyrin related glycosides and stigmasterol. The plant extract also tests positive for alkaloids (Fabio et al., 2013).

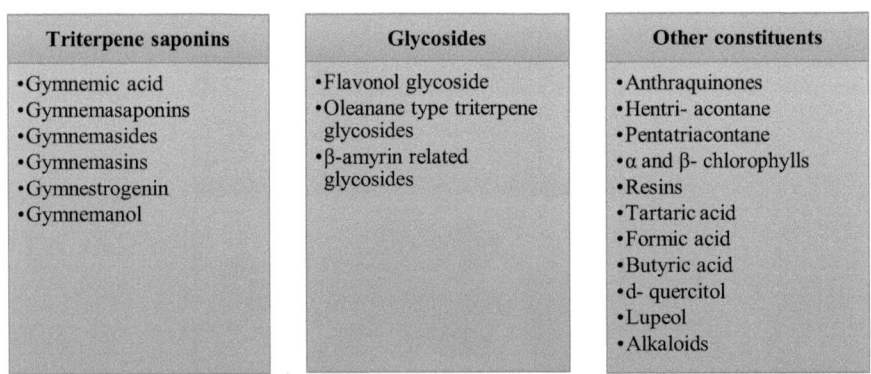

Figure: Chemical constituents of *Gymnema sylvestre* (Khramov et al., 2008; Liu et al., 2004; Fabio et al., 2013)

Table: Chemical Structure of the phytochemical constituents of *Gymnema sylvestre* (Fabio et al., 2013, Khramov et al., 2008, Liu et al., 2004)

Name of the phytoconstituents	Structure
1. Gymnemic acids	[Structure of gymnemic acids with GlcA-O, CH$_2$OH groups] |	R^1	R^2
---	---	---
Gymnemic acid I	Tygloyl	Ac
Gymnemic acid II	2-methylbutyroyl	Ac
Gymnemic acid III	2-methylbutyroyl	H
Gymnemic acid IV	Tigloyl	H	
2. Flavonol glycoside	[Structure of flavonol glycoside] R = H, R$_1$ = beta-O-Glcp-(1-->4)-alpha-L-Rhap-(1-->6)-beta-O-Galp Kaempferol 3-O-beta-D-glucopyranosyl-(1-->4)-alpha-L-rhamnopyranosyl-(1-->6)-beta-D glactopyranoside
3. Gymmestrogenin	[Structure of Gymmestrogenin]

4. Gymnemanol

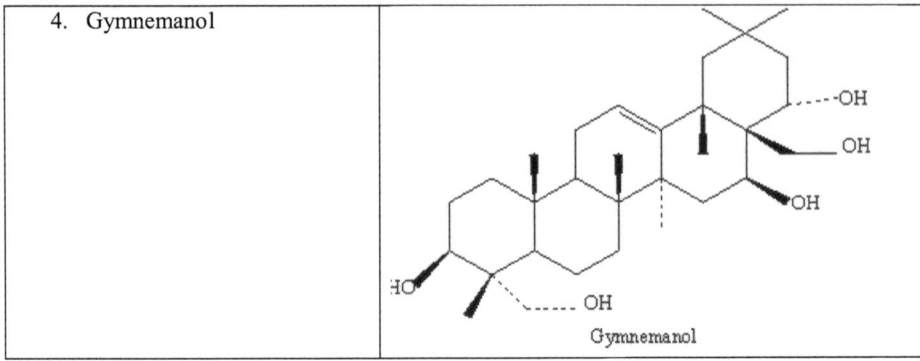

Gymnemanol

Momordica charantia: A number of reported clinical studies have shown that bitter melon extract from the fruit, seeds, and leaves contain several bioactive compounds that have hypoglycemic activity in both diabetic animals and humans (Yibchok-Anun, 2006). The main constituents that are responsible for the anti-diabetic activities are triterpene, protein, steroid, alkaloid, lipid and phenolic compounds (Saeed, 2010 & Budrat, 2008). Several glycosides have been isolated from *M. charantia* under the genera of cucurbitane- type terpenoids (Chang, 2006 & Tan, 2008).

M. charantia fruits consists glycosides, saponins, alkaloids, reducing sugars, resins, phenolic constituents, fixed oil and free acids. *M.Charantia* consists the following chemical constituents those are Alkaloids, charantin, charine, cryptoxanthin, cucurbitins, cucurbitacins, cucurbitanes, cycloartenols, diosgenin, elaeostearic acids, erythrodiol, galacturonic acids, gentisic acid, goyaglycosides, goyasaponins, guanylate cyclase inhibitors, gypsogenin, hydroxytryptamines, karounidiols, lanosterol, lauric acid, linoleic acid, linolenic acid, momorcharasides, momorcharins, momordenol, momordicilin, momordicins, momordicinin, momordicosides, momordin, momordolo, multiflorenol, myristic acid, nerolidol, oleanolic acid, oleic acid, oxalic acid, pentadecans, peptides, petroselinic acid, polypeptides, proteins, ribosome-inactivating proteins, rosmarinic acid, rubixanthin, spinasterol, steroidal glycosides, stigmasta-diols, stigmasterol, taraxerol, trehalose, trypsin inhibitors, uracil, vacine, v-insulin, verbascoside, vicine, zeatin, zeatin riboside, zeaxanthin, zeinoxanthin Amino acids-aspartic acid, serine, glutamic acid, thscinne, alanine, g-amino butyric acid and pipecolic acid, ascorbigen, b-sitosterol-d-glucoside, citrulline,
elasterol, flavochrome, lutein, lycopene, pipecolic acid. The fruit pulp has soluble pectin but no free pectic acid. Research has found that the leaves are nutritious sources of calcium,

magnesium, potassium, phosphorus and iron; both the edible fruit and the leaves are great sources of the B vitamins (Dhalla, 1969).

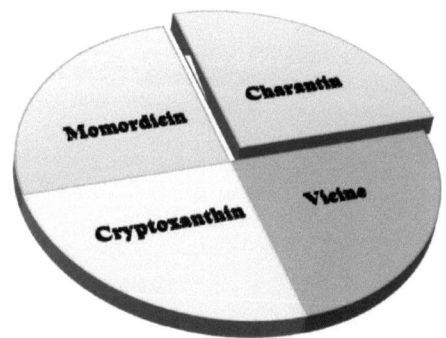

Figure: Some significant chemical constituents of *Momordica charantia* (Yibchok-Anun et al., 2006; Ghani, 1998)

Momordicin

Cryptoxanthin

Vicine

Charantin

Figure: Structures of Some Phytoconstituents Isolated From *M. charantia* L. (Yibchok-Anun et al., 2006; Ghani, 1998)

Coccinia cordifolia: *C. cordifolia* contains large amount of beta-carotene and also rich in complex carbohydrates, fibre, and a vast array of vitamins B and minerals. It is also a valuable source of nutrients. Several phytochemical constituents of great importance are reported from different parts of the *C. cordifolia* plants.

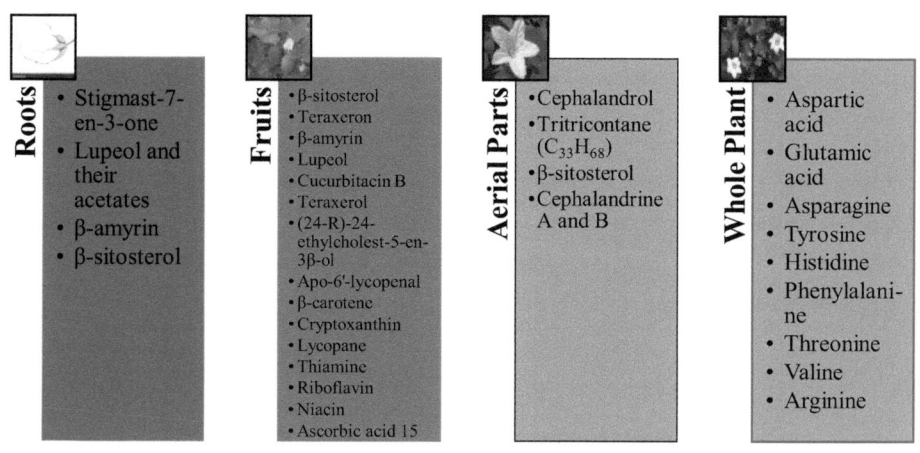

Figure: Chemical Constituents present in various parts of *Coccinia cordifolia* (Siddiqui, 1973; Orech, 2005; Umamaheswari, 2008; Syed, 2009)

The aqueous extract of fresh leaves of ivy gourd exhibited anthraquinons in addition to alkaloids, carbohydrates, proteins and amino acids, tannin, saponins, flavonoids, phytosterol, triterpenes. cephalandrol A and cephalandrol B, sigma-7-en-3-one, taraxerone and taraxerol (Rastogi, 1993). Phytochemical screening of *C. cordifolia* reported the presence of saponin, cardenoloids, flavonoids and poly phenols which may be attributed to antibacterial activity. Phenolic compounds are generally noted for their antimicrobial activities.The fruits of *C. cordifolia* are known to contain active constituents like taraxerone, taxerol, amyran, lupeol and glycoside cucurbitacan B. The leaves are bitter sweet and astringent. Major phytoconstituents present in *C. cordifolia* are cardenolides, saponins, flavonoids and polyphenols (Orech, 2005). Seed fat mainly contains palmitic (16.3%), oleic (22.4%) and linoleic (58.6%) acids. Plant also contains arabinogalactan, xyloglucan and xylan (Siddiqui, 1973).

Petroleum ether, chloroform, methanolic and aqueous extract of the leaves were tested for phytochemical analysis. The study showed that petroleum ether extract contains sterol compound and tannins, proteins and amino acid as well as glycosides were present in chloroform extract. In addition, tannins, flavonoids, glycosides, phenols, carbohydrates, proteins and amino acid, saponins and alkaloids were found in ethanolic extract whereas aqueous extract showed positive result for proteins and amino acid, glycosides, phenols, flavonoids, carbohydrates and alkaloids (Umamaheswari, 2008). Methanolic extracts of the plant also revealed the presence of alkaloids, steroids, tannins, saponins, ellagic acid, phenols, glycosides, triterpenoid and flavonoids (Syed, 2009).

Table: Major phytochemical compounds present in *Coccinia cordifolia* (Orech, 2005; Umamaheswari, 2008; Syed, 2009)

Name of the constituents	Structure
Palmitic acid	
Ascorbic acid	
Beta-amyrin	
Beta-sitosterol	
Cryptoxanthin	

Cycloartenol	
Isomultiflorenol	
Lenoleic acid	
Lupeol	
Oleic acid	
Taraxerol	

Stimast-7-en-3-one	
Thiamine	
Taraxerone	
Cucurbitacin B	

Trigonella foenum-graecum: The medicinal parts of *Trigonella foenum-graecum* are the ripe, dried seeds. Fenugreek seed contains 45-60% carbohydrates, mainly mucilaginous fiber (galactomannans); 20-30% proteins high in lysine and tryptophan; 5-10% fixed oils (lipids); pyridine-type alkaloids, mainly trigonelline (0.2-0.36%), choline (0.5%), gentianine, and carpaine; the flavonoids apigenin, luteolin, orientin, quercetin, vitexin, and isovitexin; free amino acids, such as 4-hydroxyisoleucine (0.09%); arginine, histidine, and lysine; calcium and iron; saponins (0.6-1.7%); glycosides yielding steroidal sapogenins on hydrolysis (diosgenin, yamogenin, tigogenin, neotigogenin); cholesterol and sitosterol; vitamins A, B_1, C, and nicotinic acid; and 0.015% volatile oils (n-alkanes and sesquiterpenes). The seeds also contain the saponin fenugrin B, coumarin compounds, alkaloids (trigonelline, gentianine, carpaine). A large portion of the trigonelline is degraded into nicotinic acid and pyridines, which is responsible for the flavor of the seed. The seed is also responsible for 8% of fixed, foul-smelling oil.

Several C-glycoside flavones have been identified in the seeds of fenugreek. These include vitexin, vitexin glycoside, and an arabinoside of orietin (iso-orientin), minor steroidal sapogenins (smilagenin, sarsasapogenin, yuccagenin), and up to 50% of mucilaginous fiber (Granick, 1996 & Bluementhal, 2000).

Alkaloids	Amino acids:	Saponins	Steroidal sapinogens	Flavonoids	Fibers	Other
Trimethyla-mine, Neurin, Trigonelline, Choline, Gentianine, Carpaine, Betain	Isoleucine, 4-Hydroxy isoleucine, Histidine, Leucine, lysine, L-tryptophan, Argenine	Graecunins, Fenugrin B, Fenugreeki-ne, Trigofoeno s-ides	Yamogenin Diosgenin Smilagenin Sarsasapog-enin Tigogenin Neotigogenin Gitogenin Neogitogenin Yuccagenin Saponaretin	Quercetin, Rutin, Vetixin, Isovetixin	Gum, Neutral detergent, Fiber	Coumarin, Lipid, Vitamins, Minerals, 28% mucillage, 22% proteins, 5% of stronger swelling, Bitter fixed oil

Figure: Important chemical constituents of *Trigonella feonum-graecum* (Granick, 1996; Bluementhal, 2000)

tigogenin

Gentianine

	R¹	R²
diosgenin	H	CH₃
yamogenin	CH₃	H

foenugraecin

β-D-glucose —1–6→ β-D-glucose
↑ 1–2
α-L-rhamnose

trigofoenoside A

Trigonelline

Carpaine

	R¹	R²
sarsasapogenin	CH₃	H
smilagenin	H	CH₃

Figure: Structures of Some Phytoconstituents Isolated from *Trigonella foenum-graecum* (Granick, 1996; Bluementhal, 2000).

Lagerstroemia speciosa: Leaves and fruits contain ellagitannins and related compounds. Extract of leaves contains alanine, isoleucine alpha-aminobutyric acid and methoeonine (Tanaka, 1998). Seven ellagitannins, lagerstroemin, flosin B, stachyurin , casuarinin , casuariin , epipunicacortein A , and 2, 3-(*S*)-hexahydroxydiphenoyl-α/β-d-glucose , together with one ellagic acid sulfate, 3-*O*-methyl-ellagic acid 4′-sulfate , ellagic acid , and four methyl ellagic acid derivatives, 3-*O*-methylellagic acid , 3,3′-di-*O*-methylellagic acid , 3,4,3′-tri-*O*-methylellagic acid , and 3,4,8,9,10-pentahydroxydibenzo[b,d]pyran-6-one were identified by the bioassay-directed isolation from the leaves of *Lagerstroemia speciosa* (L.) Pers. The chemical structures of these components were established on the basis of one- and two-dimensional NMR and high-resolution mass spectroscopic analyses. Other known compounds, including corosolic acid, gallic acid, 4-hydroxybenzoic acid, 3-*O*-methylprotocatechuic acid, caffeic acid, *p*-coumaric acid, kaempferol, quercetin, and isoquercitrin, were also isolated from the same plant. The obtained ellagitannins exhibited strong activities in both stimulating insulin-like glucose uptake and inhibiting adipocyte differentiation in 3T3-L1 cells. Meanwhile, ellagic acid derivatives showed an inhibitory effect on glucose transport assay. This study is the first to report an inhibitory effect for methyl ellagic acid derivatives (Bai Neisheng, 2008). All parts of the plant, particularly old leaves and ripe fruits, contain hypoglycaemic principles having activity equivalent to 6-7.7 units of insulin (Ghani, 1998). In addition, from leaves of *L. speciosa*, a new triterpenoid was isolated along with four known compounds of virgatic acid, corosolic acid, ursolic acid and β-sitosterol glucoside (Okada, 2003). Furthermore, six pentacyclic triterpenes (oleanolic acid, arjunolic acid, asiatic acid, maslinic acid, corosolic acid and 2, 3-hydroxyursolic acid) were isolated from *L. speciosa* leaves (Hou, 2007).

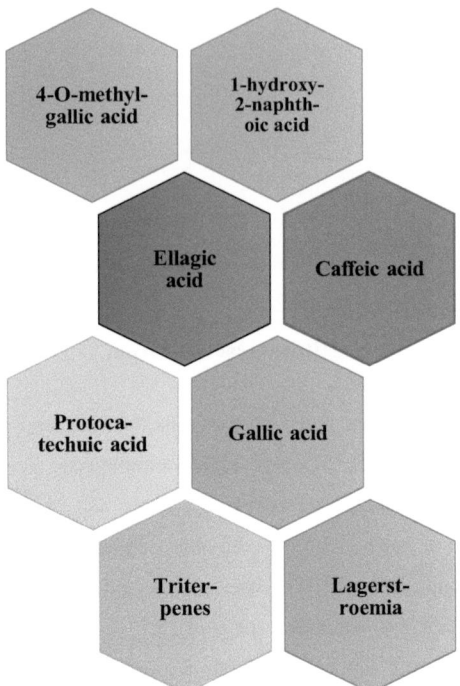

Figure: Some Phytoconstituents Isolated From *Lagerstroemia speciosa l.* (Okada, 2003; Hou, 2007; Bai Neisheng, 2008)

Table: Structures of Some Phytoconstituents Isolated From *Lagerstroemia speciosa l.* (Okada, 2003; Hou, 2007; Bai Neisheng, 2008)

Name of the constituents	Chemical Structure
1. 4-O-methylgallic acid	$R = CH_3$
2. 1-hydroxy-2-naphthoic acid	
3. Ellagic acid R1 = H, R2 = H, R3 = H (ellagic acid) R1 = CH3, R2 = H, R3 = H (3-O-methyl ellagic acid) R1 = CH3, R2 = H, R3 = CH3 (3,3'-di-O-methyl ellagic acid) R1 = CH3, R2 = CH3, R3 = CH3 (3,3',4-tri-O-methyl	
4. Caffeic acid	

5. Protocatechuic acid	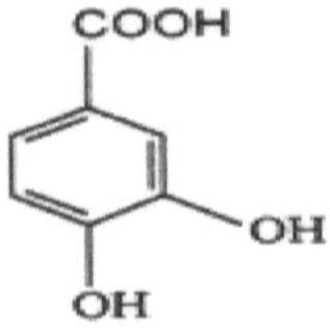
6. Gallic acid	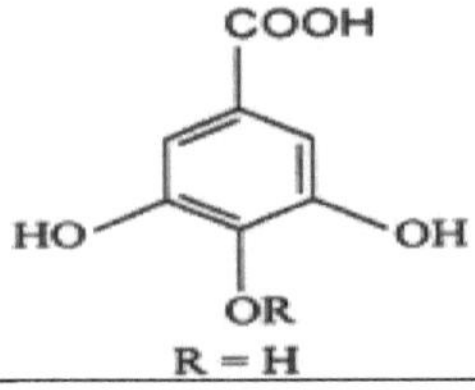
7. Triterpenes R1 = H, R2 = H (ursolic acid) R1 = OH, R2 = H (corosolic acid) R1 = OH, R2 = OH (asiatic acid)	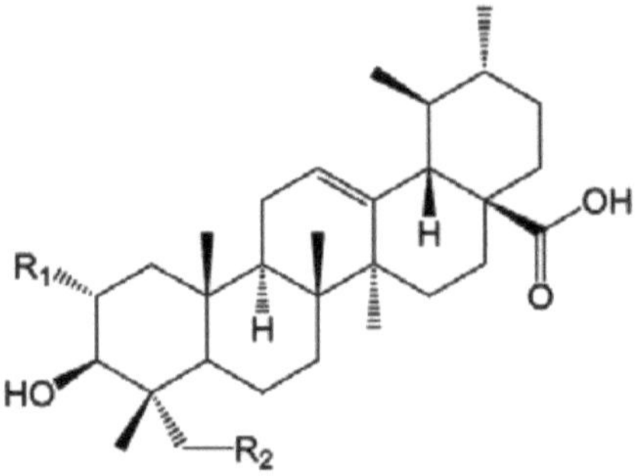
8. Lagerstroemin	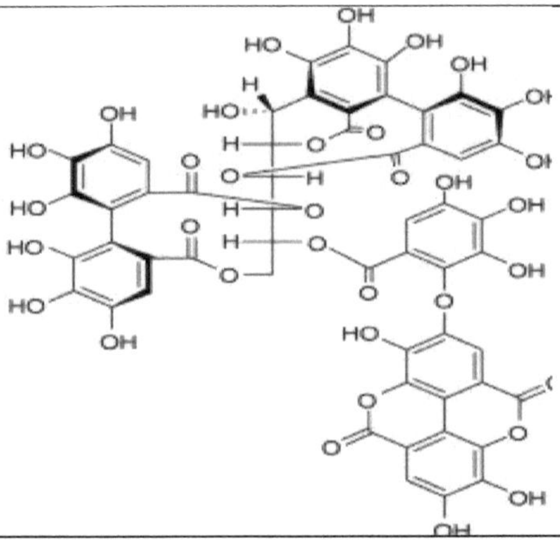

Chapter 3

- **Mechanism of action of the plants: (Kanetkar, 2007; Garau et al., 2003)**

Medicinal Plant	Mechanism of action
Gymnema sylvestre	Hypoglycemic actvivity is exhibited by means of following possible mechanisms: • Promotes regeneration of islet cells • ↑ secretion of insulin • Inhibition of glucose absorption from intestine • ↑ utilization of glucose as it ↑ the activities of enzymes responsible for utilization of glucose by insulin-dependent pathways, an increase in phosphorylase activity, decrease in gluconeogenic enzymes and sorbitol dehydrogenase (Kanetkar, 2007)
Momordica charantia	• Fruit juice ↑ glucose uptake by tissues without concomitant ↑ in tissue respiration • Aqueous extract of *M. charantia* partially stimulate insulin release from isolated beta cell of obese-hyperglycaemic mice. • Ethanolic extract of M. charantia (250 mg/kg dose orally) significantly lowered blood sugar in fasted as well as glucose loaded non-diabetic rats. • Oral administration of acetone extract of fruit powder of M. charantia for 15-30 days to alloxan-diabetic rats lowered the blood sugar and serum cholesterol levels to normal range and the blood sugar was found to be normal for up to 15 days after the end of the treatment (Garau et al., 2003)
Coccinia cordifolia	• *Coccinia cordifolia* extract could act through a variety of mechanisms including actions mimetic of those of sulfonylureas and biguanides (Akter., 2007).
Trigonella foenum-graecum	• Fenugreek has antidiabetic and antilipidemic effects. The antidiabetic effect of Fenugreek was thought to be due to formation of a colloidal-type suspension in the stomach and intestines when the mucilagenous fiber of the seeds is hydrated, therefore affecting gastrointestinal transit, slowing glucose absorption. The antilipidemic effects of Fenugreek: • inhibition of intestinal cholesterol absorption due to saponin-cholesterol complex formation, increased loss of bile through fecal excretion due to saponin-bile complexes, thus increasing conversion of cholesterol to bile by the liver, and effects of amino acid pattern of Fenugreek on serum cholesterol.
Lagerstroemia speciosa	• ↑ the secretion of insulin from pancreas and mimic the actions of sulfonylureas and miglitinides.

Chapter 4

Pharmacological Activities of Five Medicinal Plants:

Pharmacological Activities of *Gymnema sylvestre*:

Anti-diabetic Activity: The hypoglycemic action of *Gymnema sylvestre* leaves was first documented in the late 1920s. The plants hypoglycemic action is gradual in nature, differing from the rapid effect of many prescription hypoglycemic drugs. Gymnema leaves raise insulin levels by causing regeneration of Beta cells in the pancreas that secrete insulin. Research has shown that Gymnema improves the uptake of glucose into cells by increasing the activity of glucose-metabolizing enzymes, and preventing adrenaline from stimulating the liver to produce glucose, with the net effect that blood sugar levels are reduced. Another anti-diabetic effect of Gymnema is that it negates the taste of sugar, which has the effect of suppressing and neutralizing the craving for sweets. Gymnemic acid found in the leaf extracts inhibits hyperglycemia and also acts as a cardiovascular stimulant.

Oral administration of a water soluble fraction G-54 isolated from Gymnema sylvestre administered to 27 type 2 diabetic patients reduced their insulin requirement, lowered the fasting blood sugar and glycosylated haemoglobin content (Shanmugasundaram et al., 1990a). Two water soluble fractions (GS-3 and GS-4) obtained from leaves were found to double the pancreatic islets and ß-cell numbers in diabetic rats (Shanmugasundaram et al., 1990b).Alcoholic leaf extract (500mg/kg, orally) lowered maximum blood sugar in fasted, glucose fed and diabetic rats along with insulin released from pancreatic ß-cells (Chatopadhyay et al., 1993). In rats the insulin secretion from islets of Langerhans and several pancreatic ß-cell lines induced by alcoholic extract in absence of other stimulus (Persaud et al., 1999). Gymnemic acid IV, isolated from leaves produced potent hypoglycemic effect in STZ-diabetic mice (Sugihara et al., 2000). Leaf extract has been observed to produce anti-hyperglycemic (Gholap et al., 2003) and hypoglycemic (Gholap et al., 2004) effects of in corticosteroid-induced diabetes mellitus, without altered serum cortisol concentration. A polyherbal formulation containing aqueous extracts of Gymnema sylvestre produced prominent hypoglycemic activity in normal and diabetic rats at a dose of 100-500mg/kg/day, orally for acute, 6 hours and for long-term, 6 weeks studies (Mutalik et al., 2005). Gymnemic acid IV isolated from the leaves has been observed to produce hypoglycemic, anti-hyperglycemic, glucose uptake inhibitory and gut glycosidase inhibitory effects (Kimura, 2006). G. sylvestre leaves extract also treated diabetic rats' complications including hyperglycemia, hypoinsulinemia, hyperlipidemia and oxidative stress (Aziza et al., 2013).

Hypolipidemic Activity: Gymnema leaves are also noted for lowering serum cholesterol and triglycerides. Gymnema leaf extract at a dosage of 25-100 mg/kg administered orally to experimentally induced hyperlipidemic rats for two weeks reduced the elevated serum trigylceride, total cholesterol, very low density lipoprotein and low density lipoprotein-cholesterol in a dose-dependent manner. The ability of the extract at 100mg/kg to lower triglycerides and total cholesterol in serum and its anti-antheroscelrotic potential was similar to that of the standard lipid-lowering agent Clifibrate. (Bishayee and Chatterjee, 1994).

Antiobesity Study: *G. Sylvestre* helps to promote weight loss possibly through its ability to reduce cravings for sweets and control blood sugar levels. It has been reported that the gurmarin peptide block the ability to taste sweet or bitter flavors and thus reduces sweet cravings (Ninomiya et al., 1995). A standardized *G. sylvestre* extract in combination with niacin-bound chromium and hydroxycitric acid has been evaluated for antiobesity activity by monitoring changes in body weight, body mass index (BMI), appetite, lipid profiles, serum leptin and excretion of urinary fat metabolites. This study showed that the combination of *Gymnema Sylvestre* extract and hydroxycitric acid, niacin bound chromium can serve as an effective and safe weight loss formula that can facilitate a reduction in excess body weight and BMI while promoting healthy blood lipid levels (Preuss et al., 2004).

Antimicrobial Activity: The ethanolic extract of *G. sylvestre* leaves showed good antimicrobial activity against *Bacillus pumilis, B. subtilis, Pseudomonas aeruginosa* and *Staphylococcus aureus* and no activity was found against *Proteus vulgaris and Escherichia coli* (Satdive et al., 2003). The aqueous and methanolic extract of *G. sylvestre* leaves also showed moderate activity against the three pathogenic *Salmonella* species (*Salmonella typhi, S. typhimurium* and *S. paratyphi*). Out of the two extracts used, aqueous extract showed higher activity against the *Salmonella* species (Pasha et al., 2009). Ethanolic, Chloroform and Ethyl acetate extracts of the aerial parts of *G. sylvestre* also reported to have antibacterial effects against *P. vulgaris, E. coli, P. aeroginosa, Klebsella pneumoniae* and *S. aureus* (Paul et al., 2009).

Anti-Inflammatory Activity: The aqueous extract of *G. sylvestre* leaves was investigated for evaluation of anti-inflammatory activity in rats at a dose 200, 300 and 500 mg/kg in carrageenin-induced paw oedema and cotton pellet method. The aqueous extract at 300 mg/kg decreased the paw oedema volume by 48.5% within 4 h after administration, while the standard drug phenylbutazone decreased the edema volume by 57.6% when compared with the paw oedema volume of control. The aqueous extract at the dose of 200 mg/kg and 300 mg/kg produced significant reduction in granuloma weight, when compared to control group (Malik et al., 2008).

Free Radical Scavenging Activity: *In vitro*, the inhibitory effects of DPPH radicals and LDL oxidation were found with aqueous extract of *G. sylvestre*. *G. sylvestre* require 32.1 µl, for scavenging 50% of the DPPH radicals (Ohmori et al., 2005).

Effect on Taste: Gymnema or more specifically the gurmarin peptide, has been reported to block the ability to taste sweet or bitter flavors. Interference with Na+/K+ ATPase activity of taste receptors has been proposed as a possible mechanism of action. The transient effect on taste is only present when the fresh or dry leaves are chewed (Pierce, 1999).

Antiviral Activity: Gymnemic acids A, B, C and D were tested for antiviral activity against influenza virus. Gymnemic acid A (75 mg/kg/day, IP) showed the greatest activity, moderate inhibition was obtained with gymnemic acid B and none with gymnemic acids C and D.

Antibacterial Activity: Antibacterial activities of hydro-methanolic extract from leaves of *Gymnema sylvestre* was investigated using the Disk diffusion method given by Kerby-Bauer Disk Diffusion Susceptibility test. The zone of inhibition (in mm) *Gymnema sylvestre* extract exhibited strong antibacterial activities for both strain [gram (+) and gram (-) bacteria]. 50% methanolic extract of leaves of *Gymnema sylvestre* at the different concentration i.e. 25%, 50%, 75%, 100% exhibited antibacterial against and *E. coli* , Klabsella, Staphylococcus and Psuedomonas (Agrawal et al., 2016).

Antioxidant Activity: Anti-oxidant activities of *Gymnema sylvestre* extract (10-100 µg/ml) were determined according De-oxyribose method (Fenton reaction) of Halliwell and Aruoma against Ascorbic acid as standard. *In vitro* antioxidant activities of *Gymnema sylvestre* extract showed significant inhibitory concentration as compared to ascorbic acid (Agrawal et al., 2016). The active compound C-4 gem dimethyled oleanes of GS extract also possess antioxidant activity. The IC50 values for DPPH scavenging, superoxide radical scavenging, inhibition of *in vitro* lipid peroxidation, and protein carbonyl formation were 238, 140, 99, and 28 µg/mL, respectively (Sharma et al., 2009). The antioxidant activity shown by the 55% v/v alcoholic *GS* extract may be due to the presence of flavonoids, phenols, tannins and triterpenoids, all of which were detected in preliminary phytochemical screening (Yeh et al.,

2003). *In vivo* studies have shown that pre-treatment with *GS* extracts significantly enhanced the radiation (8 Gy)-induced augmentation of lipid peroxidation and depletion of glutathione and protein in mouse brain. Some multi-herbal ayurvedic formulations containing extracts of *GS*, such as Hyponidd and Dihar, have shown antioxidant activity by increasing the levels of superoxide dismutase, glutathione and catalase in rats (Babu et al., 2004; Patel et al., 2009).

Hepatoprotective Activity: An alcoholic extract of the leaf at a dose of 300 mg/kg against CCl4-induced liver damage was found to be effective.

Chemo preventive Activity: Recently, the chemo preventive activity of Gymnema sylvestre plant was successfully identified. *Gymnema sylvestre leaves* extract against 7, 12 - dimethylbenz (a) anthracene (DMBA) induced papillomagenesis in Swiss albino mice was studied. The methanolic extract of *Gymnema sylvestre* was analyzed for chemopreventive activity. Chemopreventive activity was evaluated by two stage protocol consisting of initiation with a single topical application of a carcinogen (7, 12 - dimethylbenz (a) anthracene (DMBA) followed by a promoter (croton oil) two times in a week were employed. A significant reduction in tumor incidence, tumor burden and cumulative number of papillomas was observed, along with a significant increase in average latent period in mice treated tropically with *Gymnema sylvestre extract* as compared to the control group treated with DMBA and croton oil (Agrawal et al., 2016).

Apart from all these activities, The plant is reported to be bitter, astringent, acrid, thermogenic, anti-inflammatory, anodyne, digestive, liver tonic emetic, diuretic, stomachic, stimulant, anthelmenthics, laxative, cardiotonic, expectorant, antipyretic and uterine tonic (MUM, 2003). It is useful in dyspepsia, constipation, jaundice, haemorrhoids, renal and vesical calculi, cardiopathy, asthma, bronchitis, amenorrhoea, conjuctivitis and leucoderma (Nadkarni, 1993).

Toxicity: The LD_{50} of ethanol and water extract of *Gymnema* administered intraperitoneally in mice was found to be 375 mg/kg (Bhakuni *et al*., 1971).

Pharmacological properties of *Momordica charantia*:

Anti-Diabetic Activity: Charantin (50mg/kg, orally) isolated from Momordica charantia has been resembled insulin lower blood sugar level (maximum 42% at 4th hour) of rabbits (Lolitkar et al., 1966). In a clinical study of type 1 and type 2 diabetic patients the polypeptide-p isolated from fruit, seeds and tissue exhibited hypoglycemic activity without any side effect. The subcutaneous injection of (0.5unit/kg) lowered the blood sugar in gerbils and langurs (Khanna et al., 1981). Charantin obtained from Momordica charantia induced hypoglycemic effect (Ng et al., 1986a) and also stimulated the insulin release and blocked the formation of glucose in blood stream (Ng et al., 1986b). Hypoglycemic effect and delayed cataract development was reported in alloxan diabetic rats treated with fruit extract (4g/kg/Day orally for 2 months) (Srivastava et al., 1988). Ethanolic extract (200mg/kg) of Momordica charantia was produced hypoglycemic activity in normal and streptozotocin diabetic rats; this was occurred possibly due to inhibiting glucose-6-phosphatase and fructose-1,6-biphosphatase in liver, and stimulating hepatic glucose-6- phosphate dehydrogenase activities. Oleanolic acid and momordin from plant, produced antihyperglycemic effect by inhibiting glucose transport in intestine of rat (Matsuda et al., 1988). Fruit aqueous extract (200mg/kg, orally for 6 weeks), and exercise potentially lowered blood sugar of type 2 diabetic and hyperinsulinemic (insulin resistance) rats (Miura et al., 2004). Seed aqueous extract produced prominent reduction in blood glucose, glycosylated hemoglobin, lactate dehydrogenase, glucose-6-phosphatase, fructose-1,6- bisphosphatase and glycogen phosphorylase along with increased hemoglobin, glycogen content and hexokinase, glycogen synthase activity (Sekar *et al.,* 1987). Anti-diabetic properties of plant such as charantin, vicine and polypeptide-p have the potential to be a part of dietary supplement for patients of diabetes (Krawinkel *et al.,* 2006). From *Momordica charantia* the major compounds, 5b, 19-epoxy-3b, 25dihydroxycucurbita-6, 23(E)-diene(4) and 3b-7b,25dihydroxycucurbita-5,23(E)-dien-19-al(5) administered at a dose of 400mg/kg produced hypoglycemic effect in ddY mice strain (Harinantenaina *et al.,* 2006, Gupta et al, 2008).

Cardiovascular effect: Cardiovascular effects of *Momordica charantia* was reported due to the presence of charantin. 5-10% decline in blood pressure at a dose of 800mg/kg in anaesthetized cat was determined. In addition, the contraction of isolated heart of frog was increased at dose of 5-10 mg and the same dose was effective to terminate action of acetylcholine .

Anti-sialogogue activity: Charantin at dose of 10-15 mg/kg delayed the onset of tremors but did not affect salivation produced by tremorine (Lolitkar et al, 1966).

   Lolitkar MM, Rajarama Rao MR. (1966) Pharmacology of a hypoglycaemic principles isolated from the fruits of Momordica charantia Linn. The Indian Journal of Pharmacy Vol. 28, Issue (5), pp: 129-133

Antioxidant Activities: There has been a lot of focus on the direct hypoglycemic action of bitter melon. However, some compounds have been reported to treat and prevent diabetic symptoms by mechanisms other than lowering blood glucose. For instance, many of the antioxidants found in bitter melon work by protecting the body's cells from oxidative damage. A common complication associated with diabetes is the development of atherosclerosis. This condition is caused by the build-up of reactive oxygen species (ROS) that cause lipids to be oxidized and accumulate on arterial walls, a process also known as peroxidation. These ROS are often involved in the pathogenesis of type-1 diabetes, which may account for the pancreatic damaged typically observed. One major antioxidant involved in preventing type-2 diabetes is conjugated linolenic acid (CLnA). One study discovered that CLnA occurs abundantly (57.7%) in the seed oil of bitter melon, and therefore its antioxidant activity was tested in diabetic rats. CLnA was found to significantly reduce plasma lipid peroxidation, as well as LDL-cholesterol in erythrocyte and liver tissue membranes. In addition CLnA indirectly lowers blood glucose levels since glucose auto-oxidation and protein glycation are often associated with hyperglycemia in diabetics (Dhar *et al.*, 2007).

Antiobesity Activity: Another mechanism whereby bitter melon's bioactive compounds indirectly work to lower blood glucose is by reducing adiposity and normalizing glucose tolerance (Chen et al., 2003). Obesity or excessive body fat is a major risk factor for developing type-2 diabetes. More than 80 percent of type-2 diabetics are obese or overweight. Excess fatty tissue, especially around the abdomen, causes the body's cells to become resistant to one's own insulin and leads to hyperglycemia. Therefore, many overweight diabetics can improve their blood glucose levels by losing weight through proper nutrition and regular exercise (Collazo-Clavell, 2009). Another way for diabetics to lose weight is by using herbal supplements such as bitter melon. Researchers have discovered that the bioactive compounds in bitter melon have hypolipidemic actions that can lower serum and liver cholesterol, which improves glucose tolerance. One study tested these effects on normal and STZ-induced diabetic rats fed high fat diets. Results clearly showed that bitter melon supplementation accompanied by a high fat diet reduced weight gain in the rats by preventing visceral fat accumulation. In addition, bitter melon showed marked improvements in insulin resistance and therefore glucose tolerance was normalized in the rats fed a high fat diet (Chen et al., 2003).

Hypolipidemic Activity: Compounds in bitter melon improve lipid profiles. They reduce liver secretion of apolipoprotein B (Apo B) – the primary lipoprotein of low-density "bad" cholesterol ; reduce apolipoprotein C- III expression, the protein found in very-low density cholesterol which turns into LDL/bad cholesterol; and increases the expression of apolipoprotein A-1 (ApoA1) - the major protein component of highdensity "good" cholesterol. It also lowers cellular triglyceride content. In other in vivo studies, bitter melon fruit and/or seed have been shown to reduce total cholesterol and triglycerisin both the presence and absence of dietary cholesterol. In one study, elevated cholesterol and triglyceride levels in diabetic rats were returned to normal after 10 weeks of treatment. The fruit and seed of bitter melon have demonstrated (in animal studies) to lower blood cholesterol levels. Persons on medications to lower blood cholesterol should monitor their cholesterol levels. Various cautions are indicated. (Umesh, 2005; Nerurkar, 2005)

Central Nervous System: It dissipates melancholia and gross humors [http://momordica-charantia.101herbs.com/].

Blood and Haemopoeitic Tissue: The ripe fruit of bitter melon has been shown to exhibit some remarkable anticancer effects, especially leukemia. It purifies the blood.

Anti cancer: There is absolutely no evidence that it can treat cancer. Bitter Melon and Bitter Melon Extracts inhibit cancer and tumor. A novel phytochemical in bitter melon has clinically demonstrated the ability to inhibit an enzyme named guanylate cyclase. This enzyme is thought to be linked to the pathogenesis and replication of not only psoriasis, but leukemia and cancer as well. One clinical trial found very limited evidence that bitter melon might improve immune cell function in people with cancer, but this needs to be verified and amplified in other research. Other phytochemicals that have been documented with cytotoxic activity are a group of ribosome-inactivating proteins named alpha- and beta-momorcharin, momordin,and cucurbitacin B. A chemical analog of bitter melon proteins was developed and named MAP-30 and its inventors reported that it was able to inhibit prostate tumor growth. The phytochemical momordin has clinically demonstrated cytotoxic activity against Hodgkin's lymphoma in vivo, and several other in vivo studies have demonstrated the cytostatic and antitumor activity of the entire plant of bitter melon. Further studies reported that, a water extract blocked the growth of rat prostate carcinoma and a hot water extract of the entire plant inhibited the development of mammary tumors in mice. Numerous in vitro studies have also demonstrated the anti-cancerous and anti-leukemic activity of bitter melon against numerous cell lines including liver cancer, human leukemia, melanoma and solid sarcomas (Cunnick et al, 1990; http:// www.mskcc.org/mskcc/html/69138.cfm.)

Liver and Biliary System: Fruit is useful in sub-acute cases of liver and spleen. Another method for carcinogen-induced lipid peroxidation in liver and DNA damage in lympocytes were reduced by following treatment of M.charantia. The fruit extract was found to significantly active liver enzymes glutathione stransferase, glutathione peroxidase and catalase, which showed a depression following exposure to the carcinogen. The result suggest the preventive role of water soluble constituents of M.charantia fruit during carcinogensis, which is mediated possibly by their modulatry effect on enzymes of biotransformation and detoxification system of host.

Digestive System: Leaf juice is purgative and emetic. Momordica charantia, is also a plant found in China, where it is (not surprisingly) known as Chinese Bitter Melon. It has been used in traditional Chinese medicine as an appetite stimulant, and a treatment for gastrointestinal infection.

Stomachic effect: The pure protein termed as P-insulin extracted from M. charantia fruits in crystalline form is also tested. Bitter melon contains a bitter compound called momordicin that is said to have a stomachic effect.

Skin: Fruit and leaves are used in leprosy. Bitter melon inhibits the enzyme guanylate cyclase, which may benefit people with psoriasis.

Psoriasis: A novel phytochemical in bitter melon has clinically demonstrated the ability to inhibit an enzyme named guanylate cyclase. This enzyme is thought to be linked to the pathogenesis and replication of psoriasis.

Reproductive System: leaves act as a galactogogue

Antifertility: However, toxicity and even death in laboratory animals has been reported when extracts are injected intravenously or intraperitoneally (with the fruit and seed demonstrating greater toxicity than the leaf or aerial parts of the plant). Other studies have shown ethanol and water extracts of the fruit and leaf (ingested orally) to be safe during pregnancy. The seeds, however, have demonstrated the ability to induce abortions in rats and mice, and the root has been documented with a uterine stimulant effect in animals. The fruit and leaf of bitter melon has demonstrated an in vivo antifertility effect in female animals; in male animals, it was reported to affect the production of sperm negatively. Bitter melon traditionally has been used as an abortive and has been documented with weak uterine stimulant activity; therefore, it is contraindicated during pregnancy. This plant has been documented to reduce fertility in both males and females and should therefore not be used by those undergoing fertility treatment or seeking pregnancy.

The active chemicals in bitter melon have shown in animal studies to be transferred through breast milk; therefore, it is contraindicated in women who are breast feeding.

One of the study explained that various extracts (ether, benzene and alcohol) of M. charantia seeds were administered orally and intraperitoneally to male rats for 35 days. The tests showed indirect evidence of reduced availability of pituitary gonadotrophs necessary for spermatogenesis. With intraperitoneal administration, increased cholesterol and Sudanophilic lipid levels denoted inhibited steroidogenesis, further evidence of reduced availability of gonadotrophis (Naseem et el., 1998).

Antimicrobial Agents: In addition to these properties, leaf extracts of bitter melon have clinically demonstrated broad spectrum antimicrobial activity. Various water, ethanol, and methanol extracts of the leaves have demonstrated in vitro antibacterial activities against E. coli, Staphylococcus, Pseudomonas, Salmonella, Streptobacillus and Streptococcus; an extract of the entire plant was shown to have antiprotozoal activity against Entamoeba histolytica. The fruit and fruit juice has demonstrated the same type of antibacterial properties and, in another study, a fruit extract has demonstrated activity against the stomach ulcer-causing bacteria Helicobacter pylori. Long-term use of this plant may result in the die-off of friendly bacteria with resulting yeast/candida opportunistic overgrowth. Cycling off the use of the plant (every 30 days for one week) may be warranted, and adding probiotics to the diet may be beneficial if this plant is used for longer than 30 days (Sankaranarayanan et al., 1993).

Anti-fungal Activity: Anti-fungal activity of the plant was also reported. Crude ethanolic fruit and seed extracts of *Momordica charantia* were screened against seven fungal strains. The fruit extracts of *Momordica charantia* showed moderate activity only against *A. alternata* (41.17%) and *F. oxysporum* (35.7%). Ethanolic seed extract also showed moderate activity only against *T. harizanum* (30%). Different fractions of *Momordica charantia* leaves and fruit have been previously examined for antifungal activity (Jagessar *et al.* 2008, Mwambete 2009, Santos *et al.* 2009). Antifungal ribosome inactivating proteins have also been isolated from seeds of *Momordica charantia* which are potent against *Fusarium oxysporum* and *Pythium aphanidermatum* (Wang *et al.* 2004).

Anti Viral Activity: Bitter melon (and several of its isolated phytochemicals) also has been documented with in vitro antiviral activity against numerous viruses including Epstein-Barr, herpes, and HIV viruses. In an in vivo study, a leaf extract demonstrated the ability to increase resistance to viral infections as well as to provide an immunostimulant effect in humans and animals (increasing interferon production and natural killer cell activity). Momordica Anti-human Immunovirus Protein (MAP30) activates natural killer cells, interferes with the ability of HIV viruses to divide and spread. It also increases the body's production of interferon-gamma, a natural substance that fights all types of viruses. Another clinical study showed that MAP-30's antiviral activity was also relative to the herpes virus in vitro. It contains three anti-HIV proteins: alpha- and beta momorcharin, and MAP-30, and charantin, beta-DSitosterl- beta-D-glucoside, 5,25- Stigmastadien-3-beta-D-glucoside, serotonin, and many kinds of amino acids.

Anti HIV Agents: Bitter melon has also been suggested as a treatment for AIDS, but the evidence thus far is too weak to even mention. Laboratory tests suggest that compounds in bitter melon might be effective for treating HIV infection. As most compounds isolated from bitter melon that impact

HIV have either been proteins or glycoproteins (lectins), neither of which are well-absorbed, it is unlikely that oral intake of bitter melon will slow HIV in infected people. It is possible oral ingestion of bitter melon could offset negative effects of anti-HIV drugs, if a test tube study can be shown to be applicable to people. Clearly more research is necessary before this could be recommended. The other realm showing the most promise related to bitter melon is as an immunomodulator. One clinical trial found very limited evidence that bitter melon might improve immune cell function in people with cancer, but this needs to be verified and amplified in other research. If proven correct this is another way bitter melon could help people infected with HIV. Two proteins known as alpha- and beta-momorcharin (which are present in the seeds, fruit, and leaves) have been reported to inhibit the HIV virus but research has only been demonstrated in test tubes and not in humans. Another study explained that HIV-infected cells treated with alpha- and beta-momorcharin showed a nearly complete loss of viral antigen while healthy cells were largely unaffected. "Useful for treating tumors and HIV infections, the protein is administered alone or in conjunction with conventional AIDS therapies" stated by inventors of MAP-30 protein analog in U.S. Patent. The proteins (alpha and beta momorcharin) appeared to modulate the activity of both T and B lymphocytes and significantly suppressed the macrophage activity (Bourinbaiar et el., 1995)

Larvicidal Activity: M. charantia has shown good larvicidal activity against three container breeding mosquitoes— An. stephensi, Cx. quinquefasciatus and Ae. aegypti in (Singh et el ., 2006)

Phytotoxic Activity: Crude ethanolic fruit and seed extracts of Momordica charantia were used for possible phytotoxic activity against sixteen healthy Lemna minor plants. Ethanolic fruit extract of Momordica charantia possessed highest activity (38%) at concentration of 1000 μg/ml. While ethanolic seed extracts of Momordica charantia showed low activity (18.75%) at 1000 μg/ml that is growth regulation. Previously plants like Bergenia ciliate Euphorbia wallichii and Phytolacca latbenia have been invetigated for phytotoxic activity. Phytotoxic activity has been reported for Duchesnea indica, Valeriana wallichii, Xanthium strumarium and Achyranthes aspera showing significant output. Aerial parts of Diospyros canaliculata, Paeonia emodi and Myrsine africana have been used for phytotoxic activity displaying moderate to high activities .

Anti-Genotoxic Activity: Momordica charantia decreased the genotoxic activity of methylnitrosamine, methanesulfonate and tetracycline, as shown by the decrease in chromosome breakage (Balboa et al., 1992).

Anti-helmintic activity: Momordica was more effective than piperazine in the treatment of Ascaridia galli.(Lal et al, 1976)

Wound Healing Activity: Researchers found that Momordica charantia Linn. Fruit powder, in the form of an ointment (10% w/w dried powder in simple ointment base), showed a statically significant response ($P < 0.01$), in terms of woundcontracting ability, wound closure time, period of epithelization, tensile strength of the wound and regeneration of tissues at wound site when compared with the control group, and these results were comparable to those of a reference drug povidone iodine ointment in an excision, incision and dead space wound model in rats. (Sankaranarayanan et al., 1993).

Toxicity: The seed contains vicine and therefore can trigger symptoms of favism in susceptible individuals. In addition, the red arils of the seeds are reported to be toxic to children. Many in vivo clinical studies have demonstrated the relatively low toxicity of all parts of the bitter melon plant when ingested orally.

Pharmacological Investigations on *Coccinia Cordifolia*

Anti-Diabetic Activity: Blood sugar lowering effect has been observed in patients treated with homogenized freeze dried leaves (Khan *et al.*, 1980). Ethanol extract (250mg/kg) of whole plant produced hypoglycemic activity in fasted, glucose fed and diabetic albino rats (Mukherjee *et al.*, 1988). Hypoglycemic effect of alcoholic extract (250mg/kg, orally) of *Coccinia indica* was observed in fasted and glucose fed hyperglycemic male albino rats (Chandrasekar *et al.*, 1989). Alcoholic leaf extract produced hypoglycemic effect in normal fed and 48 hours fasted rats, response mediated by suppression of gluconeogenic enzyme glucose-6-phosphatase (Hossain *et al.*, 1992). Pectin (200mg/100gm/day) isolated from fruits, exhibited blood sugar lowering effect and an increase in the glycogen content of liver in normal rats (Kumar *et al.*, 1993). Ethanol (60%) leaf extract (200mg/kg, orally) lowered the blood sugar level of diabetic rats due to suppressed glucose synthesis, through depression of glucose-6-phosphatase, fructose-1-6-biphosphatase and enhanced glucose oxidation by shunt pathway through activation of glucose-6-phosphate dehydrogenase (Shibib *et al.*, 1993). Leaf extract was produced hypoglycemic, and insulin secretogouge activity in diabetic patients (Platel *et al.*, 1997). Dried extract (500mg/kg, p.o. for 6 weeks), of plant exhibited anti-hyperglycemic activity in diabetic patients. Extract mimic insulin like activity and improved the functional status of enzymes in glycolytic pathway and lypolytic pathway (Kamble *et al.*, 1998). Potent antioxidant (Venkateswaran *et al.*, 2003) and hypolipidemic activity (Pari *et al.*, 2003) exhibited by ethanolic leaf extract administered at a dose of 200mg/kg for 45 days to streptozotocin induced diabetic rats.

Antimicrobial Activity: The bioactive compounds of fruits of *Coccinia indica* were investigated for antibacterial activity against some pathogenic bacteria. The aqueous extracts did not show much significant activity, while the organic extracts (petroleum ether and methanol) showed the highest activity against the test bacteria. The activity was more pronounced on gram-positive organisms with *staphylococcus aureus* being more susceptible and *Salmonella paratyphi a* being more resistant. Phytochemical analysis showed that the extracts contain alkaloids, tannins, saponins, flavonoids, glycosides and phenols. (Shaheen et al., 2009). *In vitro* antibacterial activity of leaves and stem extracts of *Coccinia grandis* L., has been investigated against *Bacillus cereus, Corynebacterium diptheriae, Staphylococcus aureus, Streptococcus pyogenes, Escherichia coli* (ETEC), *Klebsiella pneumonia, Proteus mirabilis, Pseudomonas aeruginosa, Salmonella typhi* and *Shigella boydii*. Water extract of leaves and ethanolic extract of stem showed significant activity against *Shigella boydii* and *Pseudomonas aeruginosa* respectively. (Farrukh et al., 2008)

Anti-Inflammatory, Analgesic and Antipyretic Activity: Both post- and pre-treatment anti-inflammatory activities of the aqueous extract of fresh leaves of *Coccinia indica* in rats were evaluated using the carrageenan-induced paw oedema method at various dose levels. Analgesic and antipyretic properties were evaluated using tail flick model and yeast-induced hyperpyrexia, respectively. Ceiling effect of the extract was observed at 50 mg/kg in pre-treatment carrageenan test. In post-treatment studies, a dose-dependent anti-inflammatory effect was observed in the dose range of 25–300 mg/kg. The effect was equivalent to diclofenac (20 mg/kg) at 50 mg/kg but it was significantly pronounced at higher doses. Effectiveness of extract in the early phase of inflammation suggests the inhibition of histamine and serotonin release. The extract produced marked analgesic activity comparable to morphine at 300 mg/kg, which suggests the involvement of central mechanisms. A significant reduction in hyperpyrexia in rats was also produced by all doses of extract with maximum effect at 300 mg/kg comparable to paracetamol. In conclusion, this study has established the anti-inflammatory activity, analgesic and antipyretic activity of *C. indica* and, thus, justifies the ethnic uses of the plant (Niazi et al .,2008).

Anti-Inflammatory and Antinociceptive Activities: The fresh fruit juice powder of *C. indica* collected in Lucknow, Uttar Pradesh, India in August 2002 was studied for its possible antiinflammatory and antinociceptive properties to rationalize the folkloric use of the plant juice as rasayana. CJP at 50-200 mg/kg significantly ($P<0.05$ to $P<0.001$) inhibited paw oedema induced by λ carrageenin (1%) and histamine (10^{-3} g/ml, 0.1 ml) in rats. The effect was comparable to the standard cycloxygenase inhibitor brufen at 100 mg/kg and the protective percentages were 63.41 and 65.78% respectively. Administration of CJP (50-200 mg/kg) moderately increased the pain threshold on analgesy-metre induced mechanical pain. However CJP significantly prevented writhing induced by acetic acid in mice and the percentages of inhibitions were 16.98-35.47%, which is equivalent to 36.67% produced by brufen. These data indicate that the fruit juice of *C. indica* rationalizes the traditional system of medicine. Again, antinociceptive activity tests were conducted in acetic acid-induced gastric pain writhing in the same mouse model. The number of writhings was induced by intraperitoneal administration of acetic acid in mice. When the lowest dose of extract tested (100 mg per kg body weight) the number of writhings was reduced by 36.4%. When a dose of 400 mg per kg body weight was given, the extract reduced the number of writhings by

47.5%. The result obtained was significantly higher when observed with a standard antinociceptive drug, aspirin (Sutradhar et al., 2011). The methanolic extract of leaf also demonstrated significant and dose-dependent antinociceptive activity.

Antioxidant Activities: The antioxidant activities of the various fractions of the hydromethanolic extract of the leaves of Coccinia grandis L. Voigt. (Cucurbitaceae) was investigatES The antioxidant activities of the fractions have been evaluated by using nine in vitro assays and were compared to standard antioxidants such as ascorbic acid, α-tocopherol, curcumin and butylated hydroxyl toluene (BHT). All the fractions showed effective H-donor activity, reducing power, free radical scavenging activity, metal chelating ability and inhibition of β-carotene bleaching. None of the fractions exerted an obvious pro-oxidant activity. The antioxidant property depends upon concentration and increased with increasing amount of the fractions. The free radical scavenging and antioxidant activities may be attributed to the presence of phenolic and flavonoid compounds present in the fractions. The results obtained in the present study indicate that the leaves of C. grandis are a potential source of natural antioxidant (*Umamaheswari & Chatterjee, 2008 ;* Upadhya et al ., 2004).

Antihyperglycemic and Antioxidative Activities: Antihyperglycemic activity study was done through oral glucose tolerance tests in glucose-loaded mice. The methanol extract of the leaf when injected to mice at doses of 50, 100, 200 and 400 mg extract per kg body weight demonstrated significant dose-dependent antihyperglycemic activity. The highest level of serum glucose reduction was observed with an extract dose of 400 mg per kg body weight, when serum glucose level was found to be reduced by 56.3%. In comparison, the standard antihyperglycemic drug, glibenclamide, when administered at a dose of 10 mg per kg body weight reduced serum glucose level in mice by 55.5%.Hence it was proved that the leaf extract of *Coccinia* has significant antihyperglycemic properties (Sutradhar et al., 2011). The antihyperglycemic and antioxidative potential of the plant Coccinia indica was assessed in alloxan induced diabetic rats. Lipid peroxidation was measured in normal, diabetic and treated animals. Blood sugar and lipid peroxidation level were higher and antioxidant level was found low in diabetic group from the normal group. A significant alteration in the blood sugar, malondialdehyde (MDA) levels and antioxidant activity was observed in diabetic animals exposed to ethanolic leaf, ethanolic leaf +alpha lipoic acid (ALA) and glibenclamide for 15 consecutive days. Coccinia indica not only reduced the oxidative stress but also strengthened the antioxidative potential (Paliwal & Khemani, 2006).

Hepatoprotective Activity: *Coccinia grandis* Linn. (Cucurbitaceae) is a perennial branched handsome tendril climber, distributed through out India. It has been used in folk medicine for the treatment of jaundice. The aim of this work was to study the hepatoprotective effect of crude ethanolic and aqueous extracts from the leaves of *C. grandis* against liver damage induced by CCl_4 in rats. The ethanolic extract at an oral dose of 200 mg kg^{-1} exhibited a significant (p<0.05) protective effect as shown by lowering serum levels of glutamic oxaloacetic transaminase, glutamic pyruvic transaminase, alkaline phosphatase, total bilirubin and total cholesterol and increasing levels of total protein and albumin levels as compared to silymarin, the positive control. These biochemical observations were supported by histopathological examination of liver sections. The activity may be due to the presence of flavonoid compounds. The extracts showed no signs of acute toxicity up to a dose level of 2000 mg kg^{-1}. Thus it could be concluded that ethanolic extract of *C. grandis* leaves possesses significant hepatoprotective activity (Sunilson et al, 2009).

Antiulcerogenic Effect: Coccinia leaves powder and methanol extract possess antiulcerogenic principle which stimulates gastric mucous secretion (Mazumder et al, 2008). Methanol extract (2g/kg), aqueous extract (2g/kg) and powder (0.5-2 g/kg) of leaves of *C. cordifolia* were tested for antiulcer activity in Wistar albino rats. Aspirin (200mg/kg bw) in 1% sodium was used as control, famotidine (20mg/kg bw) in 1% sodium was used as standard drug. Powder of leaf and methanol extract showed significant decrease of ulcer, while aqueous extract showed no significant decrease (Papiya et al., 2008). In another study ethanolic, aqueous, total aqueous extracts (200 and 400 mg/kg) of leaves of *C. cordifolia* (Linn.) were used for anti-ulcer activity. Omeprazole (2mg/kg) was used as standard drug. The ethanolic extract 400 mg/kg showed comparable anti-ulcer activity as that of standard omeprazole (Preeth et al., 2010 & Tamilselvan et al., 2011).

Antihepatotoxic Activity: Ethanolic extract of fruit and leaves of *C. cordifolia* revealed the presence of saponins. The purified fraction Ci from ethanolic extract by gradient silica gel column chromatography in the dose 25 mg/kg (Ci-1) and 50 mg/kg (Ci-2) (p.o.) showed significant dose dependent reduction in SGPT, SGOT, bilirubin, total protein, liver weight and lipid peroxide levels with reference to the standard, silymarin (25 mg/kg, p.o). The Ci compound also revealed significant dose dependent reduction in the hepatic antioxidant enzyme activities such as super oxide dismutase, glutathione, catalase, and peroxidase. The structural characterization of Ci compound by microanalysis, UV, IR, H NMR, C NMR spectroscopy and Mass spectrometry revealed structure with molecular formula C H O (beta-sitosterol). Hepatoprotective potential of Ci compound, sitosterol was inferred from its antihepatotoxic activities on serum transaminases and hepatic antioxidant enzymes in CCl₄ intoxicated rats (Gawade et al., 2012).

Antitussive activity: *C. cordifolia* has been extensively used to get relief from asthma and cough by the indigenous people of India. The antitussive effect of aerosols of two different concentrations (2.5%, 5%w/v) of methanol extract of *C. cordifolia* fruits were tested by counting the numbers of coughs produced due to aerosols of citric acid, 10 min after exposing the male guinea pigs to aerosols of test solutions for 7 min. In another set of experiment methanol extract was investigated for its therapeutic efficacy on a cough model induced by sulfur dioxide gas in mice. The results showed significant reduction of cough number obtained in the presence of both concentrations of methanol extract as compared to the prototype antitussive agent codeine phosphate. Also, methanol extract exhibited significant antitussive effect at 100, 200 and 400 mg/kg, per orally by inhibiting the cough by 20.57, 33.73 and 56.71% within 90 min of performing the experiment (Pattanayak et al., 2009). From this investigation, it can be concluded that on preliminary screening the extract of *C. cordifolia* produced a significant anti-tissue effect and thus the claim of using the plant as an anti-cough agent in ancient folklore medicine was established.

Pharmacological Investigations on *Trigonella Foenum-Graecum.L*:

Anti-Diabetic Activity: Major alkaloid trigonellin from fenugreek seeds produced hypoglycemic activity (Shani et al., 1974). Ethanol extract (0.8g/kg, i.p.) of leaves has been observed to reduce blood glucose concentration in alloxan induced diabetic rats. Lethal doses (LD50) of aqueous leaf extract were 1.9g/kg at intra-peritoneal and 10g/kg at oral administration (Abdel Barry et al., 1997). 4-Hydroxyisoleucine, an insulinotropic compound isolated from seeds increased the insulin release in glucose fed hyperglycemic rats and humans (Sauvaire et al., 1998). Seeds powder treatment normalized the enhanced lipid peroxidation and reduced the susceptibility to oxidative stress associated with depletion of antioxidants in liver of rats (Anuradha et al., 2001). Maximum 46.64% decrease in blood sugar level of diabetic rats was observed at oral administration of seed extract (1g/kg, for one month) (Vats et al., 2003). From fenugreek seeds, the soluble dietary fibre (SDF) fraction at (0.5g/kg, orally administered twice daily, for 28 days) inhibited platelets aggregation in type 2 diabetic rats and produced beneficial effect in dyslipidemia (Hannan et al., 2003). Restored activity of glutamate dehydrogenase, NAD linked isocitrate dehydrogenase and D-b-hydroxybutyrate dehydrogenase reported at oral administration of seed powder (5%, for 3 weeks) in alloxan diabetic rats. It also repaired the liver and kidney damage caused by alloxan (Thakran et al., 2004). 4-hydroxyisoleucine:5, an amino acid, isolated from seeds, produced anti-hyperglycemic effect and decreased the 33% plasma triglyceride, 22% total cholesterol (22%) and 14% free fatty acids (Narender et al., 2006 ; Gupta et al, 2008).

Analgesic and Anti-Inflammatory Activities: Analgesic and anti-inflammatory effects were examined in a partially purified fraction (MTH) of theTrigonella foenum-graecum seed extract. The analgesic effects of graded doses of fraction (MTH in 10-40 mg/kg p.o.) were evaluated in mice against acetic acid induced writhing (chemically induced pain) and hot-platemethod (thermally induced pain). The analgesia produced by MTH was compared with the standard analgesics pentazocine (PTZ, 5 mg/kg p.o.) and diclofenac sodium (DIS, 5 mg/kg p.o.). Acute anti-inflammatory activity of fraction (MTH) was also evaluated in carrageenan-induced rat paw edema model at the doses 10 and 20 mg/kg i.p. and compared with diclofenac sodium (5 mg/kg i.p.). In comparison to control group MTH showed highly significant, dose dependent analgesic activity against thermally as well as chemically induced pain ($p<0.001$). MTH at the dose of 40 mg/kg has shown significant analgesic activity ($p<0.001$) as compared to diclofenac sodium and pentazocine at the doses employed. In comparison to control, MTH at the employed doses produced marked acute anti-inflammatory activity in rats ($p < 0.001$). The results suggest that the water soluble fraction (MTH) of herbal origin has significant analgesic and anti-inflammatory potential as reflected by the parameters investigated (Vyas et al, 2008).

Effect on Blood Glucose and Lipid Profiles in Type 2 Diabetic Patients: Recently use of herbal medicines, have been considered as an alternative for therapeutic usage. So, this study was undertaken to evaluate the hypoglycemic and hypolipidemic effects of fenugreek seeds in type 2 diabetic patients.In a clinical trial study, 24 type 2 diabetic patients were placed on 10 grams/day powdered fenugreek seeds mixed with yoghurt or soaked in hot water for 8 weeks. Weight, FBS, HbA(1)C, total cholesterol, LDL, HDL and food record were measured before and after the study. The differences observed in food records, BMI and serum variables were analyzed using paired-t-test and t-student and P<or=0.05 was considered as significant. After exclusion of 6 cases for changing in medication or personal problems, the results of 18 patients (11 consumed fenugreek in hot water and 7 in yoghurt) were studied. Findings showed that FBS, TG and VLDL-C decreased significantly (25 %, 30 % and 30.6 % respectively) after taking fenugreek seed soaked in hot water whereas there were no significantly changes in lab parameters in cases consumed it mixed with yoghurt. BMI, Energy, Carbohydrate, Protein and fat intake remained unchanged during study. This study

shows that fenugreek seeds can be used as an adjuvant in the control of type 2 diabetes mellitus in the form of soaked in hot water (Kassaian et al.,2009).

Cytotoxic Activities: Cancer is the second leading cause of death worldwide. Conventional therapies cause serious side effects and, at best, merely extend the patient's lifespan by a few years. Cancer control may therefore benefit from the potential that resides in alternative therapies. There is thus an increasing demand to utilize alternative concepts or approaches to the prevention of cancer. In this report, we show a potential protective effect of Fenugreek seeds against 7,12-dimethylbenz(α)anthracene (DMBA)-induced breast cancer in rats. At 200 mg/kg b.wt., Fenugreek seeds' extract significantly inhibited the DMBA-induced mammary hyperplasia and decreased its incidence. Epidemiological studies also implicate apoptosis as a mechanism that might mediate the Fenugreek's anti-breast cancer protective effects (Amar, 2005).

Antiradical and Antioxidant Activities: An extract of fenugreek (Trigonella foenum graecum) seeds was isolated and evaluated for antioxidant activity using various in vitro assay systems. The seed extract exhibited scavenging of hydroxyl radicals (OH-) and inhibition of hydrogen peroxide-induced lipid peroxidation in rat liver mitochondria. The OH- scavenging activity of the extract was evaluated by pulse radiolysis and the deoxyribose system. The antimutagenic activity of the extract was recorded by following the inhibition of c-radiation induced strand break formation in plasmid pBR322 DNA. The extract at high concentrations acted as a scavenger of 2, 20-diphenyl-1-picryl hydrazyl hydrate (DPPH) and 2,20-azinobis 3-ethylbenzothiazoline-6-sulfonate ABTS radicals. The total phenolic content in the extract was determined spectrophotometrically according to the Folin–Ciocalteau procedure and expressed as mg or mM gallic acid equivalents. The results indicate that the extract of fenugreek seeds contains antioxidants and protects cellular structures from oxidative damage. (Kaviarasan et al., 2005)

Prophylaxis Effect: Despite considerable progress in medical therapy, there is no satisfactory drug to treat kidney stones. Therefore, the current study aimed to look for an alternative by using Trigonella foenum graecum (TFG) on nephrolithiasic rats as a preventive agent against the development of kidney stones, which is commonly used in Morocco as a phytotherapeutic agent. The inhibitory effect of the aqueous extract of TFG seeds was examined on the formation of calcium oxalate renal stones induced by ethylene glycol (EG) with ammonium chloride. At the end of the experiment all kidneys were removed and examined microscopically for possible crystal/stone locations and the total calcium amount in the renal tissue was evaluated. The blood was recovered to determine the levels of calcium, phosphorus, creatinine and urea. The results showed that the amount of calcification in the kidneys and the total calcium amount of the renal tissue in rats treated with TFG were significantly reduced compared with the untreated group (Laroubi et al.,2006).

Microdetermination of Diosgenin: Sulfuric acid hydrolysis of steroidal glycosides of Amber fenugreek was studied by capillary gas chromatographic analysis of diosgenin [(25R)-spirost-5-en-3-ol] and isomeric spirostadiene artifacts from 100 mg samples of seed material. Following extraction with 80% ethanol, highest recoveries of diosgenin occurred when hydrolyses were conducted in sulfuric acid, prepared at 1 molar (M) concentration in water containing 60−80% 2-propanol. Compared to a previous method with aqueous hydrochloric acid, the selected conditions of hydrolysis at 100 °C for 2 h with sulfuric acid in 70% 2-propanol reduced diene formation but did not completely eliminate these artifacts. Extraction of steroidal saponins with various alcohol/water mixtures prior to sulfuric acid hydrolysis gave similar recoveries of diosgenin. Application of the quantitative method to experimental samples of Amber, Quatro, and ZT-5 fenugreek, using 10 mg subsamples of crushed seed that had been defatted with petroleum ether and dried at 60 °C, gave diosgenin levels of 0.55, 0.42, and 0.75%, respectively. Levels of smilagenin and sarsasapogenin were very low in hydrolyzed seed extracts from ZT-5, a Canadian breeder line of fenugreek (Wesley G. Taylor.et al,2000).

Diuretic Activity: Trigonella foenum-Graceum Linn commonly known as fenugreek has a long history of traditional use in ayurveda and Chinese medicine and has been used for numerous indications, including labor induction, aiding digestion and to improve metabolism and diuresis. The present study was to evaluate the diuretic activity as traditionally claimed, from successive petroleum ether, benzene, ethanol and aqueous extract of fenugreek seed. The diuretic activity of the successive extract of fenugreek seeds was investigated in wistar rat, according to Lipschitz method. Phytochemical analyses of the extracts were carried out. The $Na+$, $K+$ ion concentrations were estimated by flame photometer, and $Cl-$ ion concentration was estimated by titration against silver nitrate. Results: The diuretic response and electrolyte excretion potency from petroleum ether, and benzene extract were remarkable in comparison with the control animals. The extract at 150 and 350 mg/kg body weight showed a dose dependent increase in volume of urine, the naliuretic activity seen by increase in $Na+/K+$ ions ratio with respect to control.The study indicates that aqueous and benzene extract as an effective diuretic and naluretic; thus, the work supports the traditional claim about the fenugreek seeds being used as diuretic (Das et al., 2008).

Genetic and Histopathology Studies: There is a growing interest in understanding the biological effect of medicinal plants. In the present investigation, the effects of fenugreek oil administration on the liver and ovarian activity genetically (i.e., meiotic progression in collected oocytes as well as changes in DNA and RNA content in the liver and ovarian tissues) and histopathologically (i.e., alterations in the liver and ovarian tissues) were examined in mice. Swiss albino female mice were orally administrated with different doses of fenugreek oil for 10 days. The mode and magnitude of effect were found to be depending on the dose of fenugreek oil and type of tissue. Administration with fenugreek oil at 0.1 and 0.15 ml/mouse increased the total number of cumulus-oocyte complexes as well as improved their quality. Cytogenetically, fenugreek oil was able to stimulate the oocytes collected from treated mice at all doses to progress in meiosis. Levels of nucleic acids content in all groups did not significantly change neither in the DNA nor RNA in ovarian- or liver-tissues. Histopathological examination of the ovaries collected from untreated mice as well as from mice treated with 0.05 ml/mouse of fenugreek oil showed no histopathological alterations. However, ovaries of mice treated with 0.1 or 0.15 ml/mouse of fenugreek oil showed improvement in several tissues. To our knowledge, this is the first study that suggests significant stimulating effects of fenugreek oil on the ovarian activity in mice (Hasan, 2006).

Antipyretic activity: The Assessment of antipyretic activity was carried out using Brewer's yeast induced pyrexia in Wistar rats (Loux et al., 1972). Rats were fasted overnight with water ad libitum before the experiment. The normal body temperature of each animal was measured by digital tele-thermometer and recorded. Pyrexia was induced by subcutaneously injecting 20% w/v Brewer's yeast (10 mL/kg), suspended in normal saline, into the animal's dorsum region. The peak pyrexia was observed to be at 18 h after yeast administration by conducting trial experiments. The animals that showed an increase in rectal temperature of at least 1 °C were used for the study. The drugs were administered orally at the time of peak pyrexia. The control group (group I) was administered normal saline (10ml/kg), the standard group (group II) received aspirin (100 mg/kg) and the research group (group III) was given the research drug at dose of 50 mg/kg respectively. The rectal temperature was recorded at a time interval of 1, 2, 3, 4 and 5 h after drug administration. Antipyretic effect of acetonic extract of the seeds of TFG found significant effect at the dose of 50 mg/kg in yeast-provoked elevation of body temperature when compared with control group.

Antitumor activity: A potential protective effect of Fenugreek seeds against 7,12- DMBA-induced breast cancer in rats. At 200 mg/kg b.wt., Fenugreek seeds' extract significantly inhibited the DMBA-induced mammary hyperplasia and decreased its incidence.

Epidemiological studies also implicate apoptosis as a mechanism that might mediate the Fenugreek's antibreast cancer protective effects.

Antibacterial activity: The seed extracts of Fenugreek were found moreeffective against Escherichia coli, Salmonella typhi and Staphylococus aureus.seeds were boiled in water to produce aqueous extracts.

Pharmacological Investigations on *Lagerstromia speciosa*:

Anti-Diabetic Activity: The leaves of Lagerstroemia speciosa (Lythraceae), have been traditionally consumed in various forms for treatment of diabetes and kidney related diseases. In the 1990s, the popularity of this herbal medicine began to attract the attention of scientists worldwide. Since then, researchers have conducted numerous in vitro and in vivo studies that consistently confirmed the antidiabetic activity of banaba. Scientists have identified different components of banaba to be responsible for its activity. Using tumor cells as a cell model, corosolic acid was isolated from the methanol extract of banaba and shown to be an active compound. More recently, a different cell model and the focus on the water soluble fraction of the extract led to the discovery of other compounds. The ellagitannin Lagerstroemin was identified as an effective component of the banaba extract responsible for the activity. In a different approach, using 3T3-L1 adipocytes as a cell model and a glucose uptake assay as the functional screening method, Chen et al. showed that the banaba water extract exhibited an insulin-like glucose transport inducing activity. Coupling HPLC fractionation with a glucose uptake assay, gallotannins were identified in the banaba extract as components responsible for the activity, not corosolic acid. Penta-O-galloyl-glucopyranose (PGG) was identified as the most potent gallotannin. A comparison of published data with results obtained for PGG indicates that PGG has a significantly higher glucose transport stimulatory activity than Lagerstroemin. Chen et al. have also shown that PGG exhibits anti-adipogenic properties in addition to stimulating the glucose uptake in adipocytes. The combination of glucose uptake and anti-adipogenesis activity is not found in the current insulin mimetic drugs and may indicate a great therapeutic potential of PGG (Suzuki *et al.*, 1999; Klein *et al.*, 2007).

Hypoglycemic Activity: The hypoglycemic effect of *Lagerstromia speciosa* L. leaves hot water extract on chemically induced diabetic in rat was investigated. Experimental result showed that, streptozotocin significantly ($p<0.001$) elevated the normal blood sugar level whereas treatment with hot water extract depressed the streptozotocin-induced high blood sugar level about 43.20% as compare to diabetic controls. Treatment with hot water extract increased the activity of shunt enzyme glucose–6-phosphate dehydrogenase (33.81%) and glutathione level (31.25%) and depression of the activity of hepatic gluconeogenic enzymes glucose-6-phasphatase (31.63%) and fructose–1,6-bisphosphatase (27.40%). These studies thus strongly suggest that the hot water extract of *L. speciosa* leave attributed its prominent hypoglycemic activity on experimental diabetic rats through suppression of gluconeogenesis and stimulation of glucose oxidation using the pentose phosphate pathway (Saha et al., 2009). The hypoglycemic or anti-hyperglycemic action of the leaves of *Lagerstroemia speciosa* was also reported by Garcia, 1940; Garcia, 1941; Garcia *et al.*, 1957; Kakuda *et al.*, 1996; Hayashi *et al.*, 2002; Hattori *et al.*, 2003; Liu *et al.*, 2001; Judy *et al.*, 2003; Hosoyama *et al.*, 2003; Klein *et al.*, 2007; Klein1 et al , 2007 ; Bnouham et al .,2006 ; Upadhya et al ., 2004; Tanquilut et al ., 2009)

Free Radical Scavenging: The *in vitro* antioxidant activity of the successive extracts (ethyl acetate, ethanol, methanol and water) of the leaves of *Lagerstroemia speciosa* L. (Lythraceae) were studied by examining their superoxide, hydroxyl ion scavenging and by measuring lipid peroxidation. The ethyl acetate and ethanol extracts were found to possess greater antioxidant property than the methanol and water extracts. (Priya et al., 2008).

Anti-Inflammatory Properties: Anti-inflammatory activity of the ethyl acetate and ethanol extracts were examined using the carrageenan-induced acute inflammation and formalin-induced (chronic) paw edema models. In acute and chronic inflammation models, the ethyl acetate extract reduced the paw edema significantly in a dose-dependent manner. (Priya et al., 2008).

Anti-Obesity Effect: In a study by Kakuda's research group in 1999, food containing 5% banaba water extract was used to feed female obese KK-Ay/Ta Jcl mice. Obese mice treated with banaba extract had a significantly reduced body weight (10%) compared with control mice fed with a regular diet. No change in food intake was observed. Interestingly, it was also discovered that liver triglyceride content was reduced by more than 40% in the banaba extract-treated mice. In addition, the parametrial adipose tissue was 10% lighter ($P < 0.01$) (Suzuki Y, et al, 1999).

Xanthine Oxidase Inhibitors from the Leaves of Lagerstroemia Speciosa (L.) Pers.: Xanthine oxidase (XOD) is a key enzyme playing a role in hyperuricemia, catalyzing the oxidation of hypoxanthine to xanthine and then to uric acid. This study aimed to identify the

XOD inhibitors from the leaves of Lagerstroemia speciosa (L.) Pers. (Lythraceae), which was traditionally used as a folk medicine in the Philippines. Using a bioassay-guided fractionation technique, two active compounds were isolated from the aqueous extracts of the Lagerstroemia speciosa leaves, namely valoneic acid dilactone (VAD) and ellagic acid (EA). The result demonstrated that the XOD-inhibitory effect of VAD was a stronger than that of allopurinol, a clinical drug used for XOD inhibitor, with a non-competitive mode for the enzyme with respect to xanthine as the substrate. These results may explain and support the dietary use of the aqueous extracts from Lagerstroemia speciosa leaves for the prevention and treatment of hyperuricemia (Unno et al., 2004).

Antibacterial activity: The antibacterial activity of leaves of *L. speciosa* has been reported. *L. speciosa* leaf powder extracts were tested against *Staphylococcus aureus*, *Bacillus subtilis*, *Pseudomonas aeruginosa* and *Escherichia coli* with ampicillin as standard (Ambujakshi et al., 2009). Based on the zone of inhibition, the water extract was more effective than the ethanol extract. The inhibitory efficacy of methanol extract of *L. speciosa* leaves was tested against 12 oral isolates of *Streptococcus mutans* using the agar well diffusion method (vivek et al., 2012). Results showed significant inhibitory activity against cariogenic isolates with zones ranging from 0.0–0.9 cm, 0.8–2.1 cm and 1.0–2.6 cm for extract concentrations of 10, 25 and 50 mg/ml, respectively. Flowers of *L. speciosa* have also been reported to possess antibacterial activity. Methanol extract of flower was tested against *S. mutans* and *S. aureus* using the agar well diffusion assay (Pavithra et al., 2013). The flower extract at 100 µl per well and 20 mg/ml concentration inhibited the bacteria with zones of inhibition ranging from 1.8–2.5 cm and 2.3–2.8 cm, respectively.

Antiviral activity: When tested for anti-human rhinovirus (HRV) activity in HeLa cells, orobol 7-O-D-glucoside (O7G) isolated from L. speciosa leaves showed broad-spectrum anti-HRV activity towards HRV of groups A and B (Choi et al., 2010). The inhibitory concentration (IC50) of O7G ranged from 0.58–8.80 µg/ml and the cytotoxic concentration (CC50) was more than 100 µg/ml. The compound has great potentials to be developed into a potent anti-human rhinovirus agent.

Cytotoxic activity: Using the brine shrimp (Artemia salina) lethality bioassay, the ethanol fruit extract of L. speciosa showed prominent cytotoxic activity (Rahman et al., 2011). Lethal concentration (LC50) was 60 µg/ml and LC90 was 100 µg/ml.

Anti-fibrotic activity: The effect of ethanol leaf extract of L. speciosa on male albino Wistar rats with liver fibrosis induced by carbon tetrachloride (CCl4) was studied (Prabhu et al.,

2010). Liver fibrosis was induced twice weekly by administration of CCl4 at a dose of 1 ml/kg body weight, mixed with an equal volume of corn oil. The extent of liver fibrosis was assessed by hydroxyproline content in the liver, level of aspartate transaminase, alanine transaminase, alkaline phosphatase and bilirubin in the serum, and by histological studies. Oral administration of the extract at 100 mg/kg body weight reduced the hydroxyproline content in the liver, serum enzyme levels and total bilirubin. The liver deranged by CCl4 showed improvement following administration of the extract, confirming its potent anti-fibrotic effect.

Chapter 5

Conclusion: The whole literature review emphasizes basically on the phytochemical constituents and pharmacological activities of five different medicinal plants. This extensive review provides a sharp overview about the plants and their importance to mitigate several diseases to rescue the life of the patients.

References:

A, G. (1998). Medicinal Plants of Bangladesh. *Asiatic Society of Bangladesh*, 1,7,178,305,373.

Abdel Barry, J.A., Abdel Hassan I.A. and Al-Hakiem M.H. (1997). Hypoglycaemic and antihyperglycaemic effect of *T. foenum graecum* leaf in normal and alloxan induced diabetic rats. *J. Ethnopharmacology*, **58**: 149- 155.

Ahmad, K., Shireen, F., Mehreen, & Bahar, S. (2014). Phytochemical and medicinal investigations of Momordica. VEGETOS, 27(1), 86-89. doi:10.5958/j.2229-4473.27.1.014

Agrawal, R.C, Soni, S, Jain, N , Rajpoot,J and Maheshwari,S.K (2016), Chemopreventive effect of Gymnema sylvestre in Swiss albino mice, International Journal of Scientific and Publications, Volume 6, Issue 1, page 78-83.

Akhtar MA, R. M. (2007). Comparison of long-term antihyperglycemic and hypolipidemic effects between Coccinia cordifolia (Linn.) and Catharanthus roseus (Linn.) in alloxan-induced diabetic rats. *Research Journal of Medicine and Medical Sciences, 2*, 29-34.

Amar Amin, Aysha Alkaabi, Shamaa Al-Falasi and Sayel A. Daoud (2005), Chemopreventive activities of *Trigonella foenum graecum* (Fenugreek) against breast cancer, *Cell Biology International*; **Volume 29**, Issue 8, , Pages 687-694.

Ambujakshi HR, Surendra V, Haribabu T, Goli D. Antibacterial activity of leaves of Lagerstroemia speciosa (L) Pers. Journal of Pharmacy Research 2009; 2(6):1028.

Anuradha, C.V. and Ravikumar, P. (2001). Restoration on tissue antioxidants by fenugreek seeds (*Trigonella foenum graecum*) in alloxan–diabetic rats. *Ind. J. Physiol. Pharmacol.*, **45**: 408-420.

Aziza A.M. El Shafey, Magda M. El-Ezabi, Moshira M.E. Seliem, Hannen H.M. Ouda, Doaa S. Ibrahim (2013), Effect of Gymnema sylvestre R. Br. leaves extract on certain physiological parameters of diabetic rats, Journal of King Suad University-Science, Vol. 25, Page 135-141.

Babu, P.S.; Stanely, M.P.P. Antihyperglycaemic and antioxidant effect of hyponidd, an ayurvedic herbomineral formulation in streptozotocin-induced diabetic rats. *J. Pharm. Pharmacol.* **2004**, *56*, 1435–1442.

Bai Naisheng, H. K. (2008). Active Compounds from Lagerstroemia speciosa, Insulin-like Glucose Uptake-Stimulatory/Inhibitory and Adipocyte Differentiation Inhibitory Activities in 3T3-L1 Cells. *J. Agric. Food Chem, 56*(24), 11668-11674.

Balboa JG, Lim-Sylianco CY (1992), Antigenotoxic effects of drug preparations Akapulko and Ampalaya, *Philippine Journal of Science*, **121**(4): 399.

Bhakuni D.S., Dhar M.L., Dhar M.M., Dhawan B.N (1971), Screening of Indian plants for biological activity: Part III. *Ind. J. Exp. Biol.* **9**: 91-102.

Bishayee A. and Chatterjee M (1994), Hypolipidaemic and antiatherosclerotic effect of oral *Gymnema sylvestre* R.Br. leaf extract in albino rats fed on a high fat diet. *Phytotherapy Res.* **8**(2): 118-120.

Bluementhal, G. B. (2000). Herbal medicine: expanded Commission E monographs. *Integrative Medicine Communications*, 130-133.

Bnouham Mohamed, Ziyyat Abderrahim, Mekhfi Hassane, Tahri Abdelhafid, Legssyer Abdelkhaleq (2006) Medicinal plants with potential antidiabetic activity - A review of ten years of herbal medicine research (1990-2000), *Int J Diabetes & Metabolism*; **14**: 1-25.

Bourinbaiar AS, Lee-Huang S (1995), Potentiation of anti- HIV activity of the anti-inflammatory drugsdexamethasone and indomethacin by MAP30, the antiviral agent from bitter melon. *Biochemistry and Biophysics Research Communications* **208** (2): 779.

Budrat P, S. A. (2008). Extraction of phenolic compounds from fruits of bitter melon (Momordica charantia) with subcritical water extraction and antioxidant activities of these extracts. *Chiang Mai J Sci, 35*(1), 123-130.

Chan, E. W., Tan, L. N., & Wong, S. K. (2014). Phytochemistry and Pharmacology of Lagerstroemia speciosa: A Natural Remedy for Diabetes. *International Journal of Herbal Medicine, 2*(2), 100-105.

Chandrasekar, B., Mukherjee, B. and Mukherjee, S.K. (1989). Blood sugar lowering potentiality of selected Cucurbitaceae plants of Indian origin. *Ind. J. Med. Res.*, **90**: 300-305.

Chang CI, C. C. (2006). Cucurbitane-type triterpenoids from Momordica charantia. *J Nat Prod, 71*, 1327-1330.

Chatopadhyay, R.R., Medda, C., Das, S. and Basu, T.K. (1993). Hypoglycemic and antihyperglycemic effect of *Gymnema sylvestre* leaf extract in rats. *Fitoterapia*, **64**: 450-454.

Chen, Q., Chan, L.L.Y., and Li, E.T.S (2003),. Bitter melon (*Momordica charantia*) reduces adiposity, lowers serum insulin and normalizes glucose tolerance in rats fed a high fat diet. J. Nutr.133:1088-1093.

Chodisetti, B., Rao, K., & Giri, A. (2013). Phytochemical analysis of Gymnema sylvestre and evaluation of its animicrobial activity. *Nat. Prod. Res, 27*, 583–587.

Choi HJ, Bae EY, Song JH, Baek SH, Kwon DH. Inhibitory effects of orobol 7-*O*-D-glucoside from banaba (Lagerstroemia speciosa L.) on human rhinoviruses replication. Letters in Applied Microbiology 2010; 51:1-5.

Chun, M. (2001). Biology and host specificity of Melittia oedipus (Lepidoptera: Sesiidae), a biological control agent of Coccinia grandis (Cucurbitaceae). *Proceedings of the Hawaiian Entomological Society, 35*, 85-93.

Cunnick JE, Sakamoto K, Chapes SK, Fortner GW, Takemoio DJ (1990), Induction of tumor cytotoxic immune cells using a protein from the bitter melon (Aomordica charantia). *Cellular Immunology*, **126** (2), 278.

Das Amit Kumar, Nayeem Naira, Rohini R.M (2008), Diuretic effect of Trigonella foenum graecum seed extracts, *Internet Journal of Alternative Medicine*.

Dhalla, N. G. (1961). Chemical composition of the fruit of Momordica charatia Linn. *Indian J Pharm, 23*, 128.

Dhar, P., Chattopadhyay, K., Bhattacharyya, D. Roychoudhury, A., Biswas, A., and Ghosh, S (2007), Antioxidative effect of conjugated linolenic acid in diabetic and non-diabetic blood: an *in vito* study. *J. Oleo Sci.* **56**:19-24.

Fabio, G. D., Romanucci, V., Zarrelli, M., Giordano, M., & Zarrelli, A. (2013). C-4 Gem-Dimethylated Oleanes of Gymnema sylvestre and. *Molecules, 18*, 14892-14919. doi:10.3390/molecules181214892

Farrukh Umbreen, Shareef Huma, Mahmud Shaukat, Ali Syed Ayub And Ghazala H. Rizwani (2008), Antibacterial Activities Of *Coccinia Grandis* L., *Pak. J. Bot.*, **40** (3): 1259-1262.

Flora, The Gardeners Bible. (2005). Ultimo, NSW, Australia: ABC Publishing.

G, K. (1984). *Plants for Human Consumption.* Koenigstein, Germany: Koeltz Scientific Books.

Garau, C., E, C., Phoenix, D., & Singh, J. (2003). Beneficial effect and mechanism of action of Momordica charantia in the treatment of diabetes mellitus: a mini review. *Int J Diabetes & Metabolism, 11*, 46-55.

Garcia F (1940). On the hypoglycemic effect of decoction of *Lagerstroemia speciosa* leaves (banaba) administered orally. *J Phil Med Assoc.* **20**: 395–402.

Garcia F (1941), Distribution and deterioration of insulin-like principle in *Lagerstroemia speciosa* (banaba). *Acta Med Philippina.* **3**: 99–104.

Garcia F,. Melencio-Maglalang P (1957), Application of banabins (a plantisul preparation) and S.B. menus to diabetics. *J Phil Med Assoc.* **33**(1): 7–15.

Gautam, S., Meshram, A., & Srivastava, N. (2014). A BRIEF STUDY ON PHYTOCHEMICAL COMPOUNDS PRESENT IN COCCINIA CORDIFOLIA FOR THEIR MEDICINAL, PHARMACOLOGICAL AND INDUSTRIAL APPLICATIONS. WORLD JOURNAL OF PHARMACY AND PHARMACEUTICAL SCIENCES, 3(2).

Gawade SP, Rao MVC. (Antihepatotoxic Activities of Ci Compound: ß Sitosterol Isolated from Fruits and Leaves of *Coccinia indica*). Ind J Pharm Edu Res, 2012; 46(1) :7-11.

George M., Joseph L., Sabu NS., Evaluation of antipyretic and anti-inflammatory activity of seed extract of Trigonella foenum graecum, Journal of Chemical and Pharmaceutical Research 8(1):132-135.

Ghani, A. (1998). Medicinal Plants of Bangladesh. *Asiatic Society of Bangladesh*, 1,7,178,305,373.

Gholap, S. and Kar, A. (2003). Effects of *Inula racemosa* root and *Gymnema sylyestre* leaf extracts in the regulation of corticosteroid induced diabetes mellitus: involvement of thyroid hormones. *Pharmazie*, **58**: 413- 415.

Gholap, S. and Kar, A. (2004). Hypoglycaemic effects of some plant extracts are possibly mediated through inhibition in corticosteroid concentration. *Pharmazie,* **59**: 876-878.

Granick B, N. D. (1996). The Lawrance review of natural products. *St. Louis: Facts and Comparisons*, 1-3.

Gupta BD, Dandiya PC, Gupta ML. (1971) A psychopharmacological analysis of behavior in rat. *Jpn J Pharmacol.* **21**:293

Gurav S., G. V. (2007). *Pharmacog rev, 1*(2), 338-343.

Hannan, J.M., Rokeya, B., Faruque, O., Nahar, N., Mosihuzzaman, M., Azad Khan, A.K. and Ali, L. (2003). Effect of soluble dietary fibre fraction of *Trigonella foenum graecum* on glycemic, insulinemic, lipidemic and platelet aggregation status of Type 2 diabetic model rats. *J. Ethnopharmacology*, **88**: 73-77.

Harinantenaina, L., Tanaka, M., Takaoka, S., Oda, M., Mogami, O., Uchida, M. and Asakawa, Y. (2006). *Momordica charantia* constituents and anti-diabetic screening of the isolated major compounds. *Chem. Pharm. Bull.*, **54**: 1017-1021.

Hassan, Aziza M., Wagdy K.B. Khalil and Kawkab A. Ahmed (2006), Genetic and histopathology studies on mice: effect of fenugreek oil on the efficiency of ovarian and liver tissues, *African Journal of Biotechnology* **5**, 5, 477—483.

Hattori K, Sukenobu N, Sasaki T, Takasuga S, Hayashi T, Kasai R (2003), Activation of insulin receptors by lagerstroemin. *J Pharmacol Sci.* **93**(1): 69–73.

Hayashi T, Maruyama H, Kasai R, Hattori K, Takasuga S, Hazeki O (2002). Ellagitannins from Lagerstroemia speciosa as activators of glucose transport in fat cells. Planta Me. 68(2): 1735.

Hosoyama H, Sugimoto A, Suzuki Y, Sakane I, Kakuda T (2003). Isolation and quantitative analysis of the alpha-amylase inhibitor in *Lagerstroemia speciosa*(L.) Pers. (Banaba). *Yakugaku Zasshi.* **123**(7):599–605.

Hossain, M.Z., Shibib, B.A. and Rahman, R. (1992). Hypoglycemic effects of *Coccinia indica*: Inhibition of key gluconeogenic enzyme, glucose-6- phosphatase. *Ind. J. Exp. Biol.*, **30**: 418-420.

Hou W, L. Y. (2009). Triterpene acids isolated from Lagerstroemia speciosa leaves as α-glucosidase inhibitors. *Phytotherapy Research, 23*, 614-618.

Jagessar RC and Mohameda G (2008) An evaluation of the Antibacterial and Antifungal activity of leaf extracts of Momordica Charantia against Candida albicans, Staphylococcus aureus and Escherichia coli. Nature and Science 6(1): 1545-0740.

JJ Loux,PD Depalma,SL Yankell. Toxicol Appl Pharmacology,1972, 22(4),672–675.

Judy WV, Hari SP, Stogsdill WW, Judy JS, Naguib YM, Passwater R (2003), Antidiabetic Activity of a standardized extract (Glucosol) from *Lagerstroemia speciosa* leaves in type II diabetics. A dose-dependence study. *J Ethnopharmacol.* **87**(1):115–7.

Kakuda T, Sakane I, Takihara T, Ozaki Y, Takeuchi H, Kuroyanagi M.(1996), Hypoglycemic effect of ex extracts from *Lagerstroemia speciosa* L. leaves in genetically diabetic KK-AY mice. *Biosci Biotechnol Biochem.* **60**(2): 204–8.

Kamble, S.M., Kamlakar, P.L., Vaidya, S. and Bambole, V.D. (1998). Influence of *Coccinia indica* on certain enzymes in glycolytic and lipolytic pathway in human diabetes. *Ind. J. Med. Sc.*, **52**:143-146.

Kanetkar P., S. R. (2007). *J Clin Biochem Nutr, 41*, 77-81.

Kassaian N, Azadbakht L, Forghani B, Amini M (2009), Effect of fenugreek seeds on blood glucose and lipid profiles in type 2 diabetic patients. *Int J Vitam Nutr* Res.; **79**(1):34-9.

Kaviarasan Kaviarasan Keshavamurthy K.R., Y. S. (1990). *Flora of Coorg- Karnataka*. Banglore: Vimsat publishers.

Keshavamurthy K.R., Yoganarasimhan S.N (1990), Flora of Coorg – Karnataka, Vimsat publishers, Banglore, 282.

Khan, A.K.A., Akhtar, S. and Mahatab, H. (1980). Treatment of diabetes mellitus with *Coccinia indica*. *Brit. Med. J*; **280**(6220): 1044.

Khanna, P., Jain, S.C., Panagariya, A. and Dixit, V.P. (1981). Hypoglycemic activity of polypeptide-p from a plant source. *J. Nat. Prod.*, **44**: 648-655.

Khramov V.A., S. A. (2008). *Pharm Chem J, 42*(1), 29-31.

Klein G, Kim J, Himmeldirk K, Cao Y and Chen X. 2007. Antidiabetes and Anti-obesity Activity of *Lagerstroemia* speciosa. *eCAM*. **4**(4): 401-407.

Klein G, Kim J, Himmeldirk K, Cao Y and Chen X. 2007. Antidiabetes and Anti-obesity Activity of *Lagerstroemia* speciosa. *eCAM*. **4**(4): 401-407.

Krawinkel, M.B. and Keding, G.B. (2006). Bitter gourd (*Momordica charantia*): *A dietary approach to hyperglycemia. Nutr.* Rev., **64**: 331-337.

Kumar, A. (2016). Gymnema Sylvestre: Qualitative Analysis of Phytochemicals and its Medicinal Importance in Different Medicinal System. *Asian Journal of Biochemical and Pharmaceutical Research,* 6(1), 2231-2560.

Kumar, G.P., Sudheesh. S. and Vijayalakshmi, N.R. (1993). Hypoglycemic effect of *Coccinia indica*:Mechanism of action. *Planta Medica*, **59**: 330-332.

Kunkel, G. (1984). *Plants for Human Consumption*. Koenigstein, Germany: Koeltz Scientific Books.

Kuzuko Y, K. A. (1989). Structure studies of new antisweet constituents from Gymnema sylvestris. *Tetrahedron lett, 30*(9), 1130-1106.

Lal J, Chandra S, Raviprakash V, Sabir M (1976), In vitro anthelmintic action of some indigenous medicinal plants on Ascaridia galli worms. *Indian Journal of Physiology and Pharmacology* **20** (2): 64.

Laroubi Amine; Touhami Mohammed ; Farouk Loubna ; Zrara Ibtissam ; Aboufatima Rachida ; Benharref Ahmmed ; Chait Abderrahman (2007), Prophylaxis Effect Of Trigonella Foenum Graecum L. Seeds On Renal Stone Formation In Rats , *Wiley InterScience* ; **Vol. 21**(10), pp. 921-925.

Liu F, Kim J, Li Y, Liu X, Li J, Chen X (2001), An extract of *Lagerstroemia speciosa* L. has insulin-like glucose uptake-stimulatory and adipocyte differentiation-inhibitory activities in 3T3-L1 cells. *J Nutr.* **131**(9): 2242–7.

Liu X., Y. W. (2004). *Carb Res, 339*(4), 891-95.

Lolitkar, M.M., Rao, M.R.R. (1966). Pharmacology of a hypoglycemic principle isolate from fruit of *Momordica charantia* Linn. *Ind. J. Pharmacy*, **28**:129-133.

Madhurima, Ansari, S., Alam.P., & Ahmed, S. M. (2009). Pharmacognostic and phytochemical analysis of Gymnema sylvestre R (Br) leaves. *Journal of Herbal medicine and Toxicology, 3*(1), 73-80.

MA, Uddin N, Hasanuzzaman M, Rahman AA. Antinociceptive, antidiarrhoeal and cytotoxic activities of *Lagerstroemia speciosa* (L.) Pers. Pharmacologyonline 2011; 1.

Malik J.K., Manvi F.V., Alagawadi K.R, Noolvi M (2008), *Int J Green Pharm*, **2**(2), 114-15.

Manandhar, N. (2002). *Plants and people of Nepal*. Timber Press Oregon.

Mazumder Papiya mitra, susmal D and nimbi arivudai R, (2007), Antiulcerogenic and antioxidant effect s of coccinia grandis (Linn) voigt leaves on aspirin – induced ulcer in rat, Natural product radiance, vol 7(1) ,2008 ,pp 15-18.

ME, C. (2001). Biology and host specificity of Melittia oedipus (Lepidoptera: Sesiidae), a biological control agent of Coccinia grandis (Cucurbitaceae). *Proceedings of the Hawaiian Entomological Society, 35*, 85-93.

Miura, T., Itoh, Y., Iwamoto, N., Kato, M. and Ishida, T. (2004). Suppressive activity of the fruit of *Momordica charantia* with exercise on blood glucose in type 2 diabetic mice. *Biol. Pharm. Bull.*, **27**: 248-250.

Mukherjee, B., Sekar, B.C. and Mukherjee, S.K. (1988). Blood Sugar lowering effect of *Coccinia indica* root and whole plant, in different experimental rat models. *Phytoterapia*, **59**: 207- Rahman

Mwambete K (2009) The *in vitro* antimicrobial activity of fruit and leaf crude extracts of *Momordica charantia:* A Tanzania medicinal plant. Afr Health Sci 9(1): 34–39.
612.

MUM (2003), *Monographs of Unani Medicine, Vol-1*. Bait al-Hikmah, Hamdard Foundation Pakistan, Islamabad, Pakistan. pp. 271-272.

Mutalik, S., Chetana, M., Sulochana, B., Devi, P.U. and Udupa, N. (2005). Effect of Dianex, an herbal formulation on experimentally induced diabetes mellitus. *Phytotherapy Research*, **19**: 409- 415.

Nadkarni K.M (1993), *Indian Materia Medica*, Popular prakashan, Bombay. **1**: 596-599.

Narender, T., Puri, A., Shweta, Khaliq, T., Saxena, R., Bhatia, G. and Chandra, R. (2006). 4-hydroxyisoleucine an unusual amino acid as antidyslipidemic and antihyperglycemic agent. *Bioorg. Med. Chem. Lett.*, **16**: 293- 296.

Naseem Mulla Zehera, Patil Srinivas Reddy, Patil Somnath Reddy, Ravindra and Patil Saraswati B (1998), Antispermatogenic and androgenic activities of Momordica charantia (Karela) in albino rats. *J Ethnopharmacology*, **61**: 9-16.

Natural Medicines Comprehensive Database Web site. Available at: http://www.naturaldatabase.com/monograph.asp?mono_id=733&hilite=1 Accessed January 29, 2003.

Nerurkar Pratibha V, Laurel Pearson, Jimmy T. Efird, Khosrow Adeli, Andre G. Theriault and Vivek R. Nerurkar (2005), "Microsomal Triglyceride Transfer Protein Gene Expression and ApoB Secretion are Inhibited by Bitter Melon in HepG2 Cells" *J.Nutr*, **135**(4): 702-706.

Ng, T.B., Wong, C.M., Li, W.W. and Yeung, H.W. (1986a). Insulin-like molecules in *Momordica charantia* seeds. *J. Ethnopharmacology*, **15**: 107-117.

Ng, T.B., Wong, C.M., Li, W.W. and Yeung, H.W. (1986b). Isolation and Characterization of a galactose binding lectin with insulinomimetic activities from the seeds of the bitter gourd *Momordica charantia* (Family: Cucurbitaceae). *Intrnational J. Peptide Protein Research*, **28**: 163-172.

Niazi Junaid, Singh Parabhdeep, Bansal Yogita and Goel R. K. (2009), Anti-inflammatory, analgesic and antipyretic activity of aqueous extract of fresh leaves of *Coccinia indica, Inflammopharmacology*, **Volume 17**(4),pp 239-244.

Ninomiya Y., Imoto T (1995), *Am J Physiology*, **268** (4), 1019-25.

NP, M. (2002). *Plants and people of Nepal.* Timber Press Oregon.

Ohmori R, Iwamoto T, Tago M, Takeo T., Unno T., Itakura H (2005), *Lipids*, **40** (8), 849-53.

Okada Y, O. A. (2003). A new triterpenoid isolated from Lagerstroemia speciosa (L.) Pers. *Chemical and pharmaceutical Bulletin, 51*(4), 452-454.

Orech FO, A. T.-H. (2005). Potential toxicity of some traditional leafy vegetables consumed in Nyang'oma division, western Kenya. *Afr J Food and Nutritional S, 5*, 1-13.

Paliwal Pritee, Khemani L.D (2006), Antihyperglycemic and antioxidative activities of *Coccinia indica* in alloxan induced diabetic rats, *The Indian Journal of Veterinary Research;* **Vol : 15**(2).

Papiya MM, Sasmal D, Arivudi NR. (Antiulcerogenic and antioxidant effects of *Coccinia grandis* (Linn). Voigt leaves on aspirin-induced gastric ulcer in rats). Nat Prod Radiance, 2008; 7(1): 15-18.

Pari, L. and Venkateswaran, S. (2003). Protective effect of *Coccinia indica* on changes in the fatty acid composition in streptozotocin induced diabetic rats. *Pharmazie*, **58**: 409-412.

Pasha C.; Sayeed S.; Ali S.; Khan Z (2009), *Turk J Biol*, **33**, 59-64.

Patel, S.S.; Shah, R.S.; Goyal, R.K. Antihyperglycemic, antihyperlipidemic and antioxidant effects of dihar, a polyherbal ayurvedic formulation in streptozotocin induced diabetic rats. Indian J. Exp. Biol. 2009, 47, 564570.

Pattanayak PS, Priyashree S. (In vivo antitussive activity of *Coccinia grandis* against irritant aerosol and sulfur dioxide induced cough model in rodents). Bangladesh J Pharmacol, 2009; 4: 84-7.

Paul J.P., Jayapriya K (2009), *Pharmacologyonline*, **3**, 832-836.

Pavithra GM, Rakesh KN, Dileep N, Syed Junaid, Ramesh Kumar KA, Prashith Kekuda TR. Elemental analysis, antimicrobial and radical scavenging activity of *Lagerstroemia speciosa (L.) flower. Journal of Chemical and Pharmaceutical Research 2013; 5(6):215-222.*

Persaud, S.J., Al-Majed, H., Raman, A. and Jones, P.M. (1999). *Gymnema sylvestre* stimulates insulin release in vitro by increased membrane permeability. *J. Endocrinology*, **163**: 207-212.

Pierce A (1999), Gymnema Monograph. Practical guide to natural medicine. New York: Stonesong Press Book; p. 324-26

Platel, K. and Srinivasan, K. (1997). Plant foods in the management of diabetes mellitus: vegetables as potential hypoglycemic agents. Die. *Nahrung,* **41**: 68-74.

Potawale E., S. V. (2008). *Pharmacology online, 2,* 144-157.

Prabhu VV, Chidambaranathan N, Nalini G, Venkataraman S, Jayaprakash S, Nagarajan M. Evaluation of anti-fibrotic effect of *Lagerstroemia speciosa* (L) Person carbon tetrachloride induced liver fibrosis. Current Pharmaceutical Research 2010; 1(1):7-12.

Preeth M, Shobana J, Upendarrao G , Thangathirupathi A. (Antiulcer effect of *Coccinia grandis (Linn.) on pylorus ligated (albino) rats). Int J Pharm Res Dev, 2010; 2(5): 1-9.*

Priya T. T. , Sabu M. C. and Jolly C. I (2008), Free radical scavenging and anti-inflammatory properties of *Lagerstroemia speciosa* (L), *Inflammopharmacology*; **Vol 16**(4) pp182-187.

Ramachandaran K, S. B. (1983). (Scarlet gourd, Coccinia indica, little known tropical drug plants). *Econ Bot, 34*(4), 380-383.

Rastogi RP, M. B. (1993). *Compendium of Indian Medicinal Plants,.* New Delhi: CDRI, Lucknow, Publication and Information Directorate.

Retrieved 11 ,10, 2016, from http://www.bitterrootrestoration.com/medicinal-plants/bitter-melon.html.

Retrieved from http://www.bitterrootrestoration.com/medicinal-plants/bitter-melon.html.

Saeed MK, S. I. (2010). Nutritional analysis and antioxidant activity of bitter gourd (Momordica charantia) from Pakistan. *Pharmacologyonline, 1,* 252-260.

Saho N.P., M. S. (1996). *Phytochemistry, 41*(4), 1181-85.

Sankaranarayanan J, Jolly CI (1993), Phytochemical, antibacterial, and pharmacological investigations on Momordica charantia Linn., Emblica offidnalis Gaertn. and Curcuma longa Linn. *Indian Journal of Pharmaceutical Science* **55**(1): 6.

Santos KK, Matias EF, Sobral CE, Tintino SR, Morais MF, Guedes GM, Santos FA, Sousa AC,

Rolón M, Vega C, Arias AR, Costa JG, Menezes IR and Coutinho HD (2009) Trypanocide, cytotoxic, and antifungal activities of Momordica charantia. Pharm Biol 2: 162-6.

Satdive.K., Abhilash P., Devanand P.F(2003), *Fitoterapia*, **74**, 699-701.

Sauvaire, Y., Petie, P., Broca, C., Manteghetti, M., Baissac, Y., Fernandez-Alvarez, J., Gross, R., Roye, M., Leconte, A., Gomis, R. and Ribes, G. (1998). 4-Hydroxyisoleucine: a novel amino acid potentiator of insulin secretion. Diabetes, **47**: 206-210.

Sekar, D.S., Sivagnanam, K. and Subramanian, S. (2005). Hypoglycemic activity of *Momordica charantia* seeds on streptozotocin induced diabetic rats. *Pharmazie*, **60**: 283-387.

Shaheen Syed Zeenat, Bolla Krishna, Vasu Kandukuri and Charya Singara M. A. (2009), Antimicrobial activity of the fruit extracts of *Coccinia indica, African Journal of Biotechnology*; **Vol. 8** (24), pp. 7073-7076.

Shanmugasundaram, E.R.B., Rajeswari, G., Bhaskaran, K., Rajesh Kumar, B.R., Raja, K. and Kijar Ahmad (1990a), Use of *Gymnema sylvestre* leaf extract in the control of blood glucose in insulin dependent diabetes mellitus. *J. Ethnopharmacology*, **30**: 281-294.

Shanmugasundaram, E.R.B., Gopinath, K.L., Shanmugasundaram, K.R. and Rajendran, V.M. (1990b). Possible regeneration of islets of langerhans in streptozotocin-diabetic rats given *Gymnema sylvestre* leaf extracts. *J. Ethnopharmacology*, **30**: 265-279.

Sharma, K.; Singh, U.; Vats, S.; Priyadarsini, K.; Bhatia, A.; Kamal, R. Evaluation of evidenced-based radioprotective efficacy of *Gymnema sylvestre* leaves in mice brain. *J. Environ. Pathol. Toxicol. Oncol.* **2009**, *28*, 311–323.

Siddiqui I A, O. S. (1973). Fatty acid composition of seed fats from four plant families. *J Oil Technol Assoc India, 5*, 8-9.

Singh R.K., Dhiman R.C. & Mittal P.K (2006), Mosquito larvicidal properties of Momordica charantia Linn (Family: Cucurbitaceae) *J Vect Borne Dis* **43**, pp. 88–91.

Srivastava, Y., Venkatakrishna-Bhatt, H. and Verma, Y. (1988). Effect of *Momordica charantia* Linn. pomous aqueous extract on cataractogenesis in murrin alloxan diabetics. *Pharmacol. Res. Commun.*, **20**: 201- 209.

Sutradhar BK, Islam MJ, Shoyeb MA, Khaleque HN, Sintaha M, Noor FA, Walid N, Mohammed R. (An Evaluation of Antihyperglycemic and Antinociceptive Effects of Crude Methanol Extract of *Coccinia grandis* (L.) J. Voigt. (Cucurbitaceae) Leaves in Swiss Albino Mice). Adv. in Nat. Appl Sci, 2011; 5(1) : 1-5.

Sunilson, J.A.J., M. Muthappan, A. Das, R. Suraj, R. Varatharajan and P. Promwichit (2009), Hepatoprotective activity of *Coccinia grandis* leaves against carbon tetrachloride induced hepatic injury in rats. *Int. J. Pharmacol.*, **5**: 222-227.

Suzuki Y, Unno T, Ushitani M, Hayashi K, Kakuda T. (1999) Antiobesity activity of extracts from Lagerstroemia speciosa L. leaves on female KK-Ay mice. *J Nutr Sci Vitaminol* **45**::6791–5.

Syed Z, K. B. (2009). Antimicrobial activity of the fruit extracts of Coccinia indica. *Afr J Biotechnol, 8*(24), 7073-76.

Tamilselvan N, Thirumalai T, Elumalai EK, Balaji R, David E. (Pharmacognosy of *Coccinia grandis*: a review). Asian Pac J Trop Biomed, 2011; S299-S302.

Tan MJ, Y. J.-B. (2008). Antidiabetic activities of triterpenoids isolated from bitter melon associated with activation of the AMPK pathway. *Chem Biol, 15*, 263-273.

Tanaka, T. T. (1992). *40*(11).

Tanquilut N. C., Tanquilut M. R. C., Estacio M. A. C., Torres E. B., Rosario J. C. and Reyes B. A. S (2009), Hypoglycemic effect of *Lagerstroemia speciosa* (L.) Pers. on alloxan-induced diabetic mice, *Journal of Medicinal Plants Research* **Vol. 3**(12), pp. 1066-1071.

Thakran, S., Siddiqui, M.R. and Baquer, N.Z. (2004). *Trigonella foenum graecum* seed powder protects against histopahological abnormalities in tissues of diabetic rats. *Mole. Cellular Biochem.*, **266**: 151-159.

Umamaheswari M., Chatterjee T. K. (2008), In Vitro Antioxidant Activities Of The Fractions Of Coccinia Grandis L. Leaf Extract, *African Journal Of Traditional, Complementary And Alternative Medicines (AJTCAM);* **5**(1): 61-73

Umesh C. S. Yadav, K. Moorthy, Najma Z. Baquer (2005), Combined treatment of sodium orthovanadate and Momordica charantia fruit extract prevents alterations in lipid profile and lipogenic enzymes in alloxan diabetic rats, *Molecular and Cellular Biochemistry*, **268**(1-2), 111-120.

Unno T, Sugimoto A, Kakuda T(2004), Xanthine oxidase inhibitors from the leaves of Lagerstroemia speciosa (L.) Pers., *J Ethnopharmacol*; 93(2-3):391-5.

Vats, V., Yadav, S.P. and Grover, J.K. (2003). Effect of *T. foenum graecum* on glycogen content of tissue and the key enzymes of carbohydrate metabolism. *J. Ethnopharmacology*, **85**: 237-242.

Venkateswaran, S. and Pari, L. (2003b). Effect of *Coccinia indica* leaves on antioxidant status in streptozotocin–induced diabetic rats. *J. Ethnopharmacology*, **84**: 163-168.

Vivek MN, Sunil SV, Pramod NJ, Prashith Kekuda TR, Mukunda S, Mallikarjun N. Anti cariogenic activity of *Lagerstroemia speciosa* (L.). Science, Technology and Arts Research Journal 2012; 1(1):53-56.

Vyas S, Agrawal RP, Solanki P, Trivedi P (2008), Analgesic and anti-inflammatory activities of Trigonella foenum-graecum (seed). *Acta Pol Pharm*.; 65(4):473-6.

Wani, S. A., & Kumar, P. (2016). Fenugreek: A review on its nutraceutical properties. *Journal of the Saudi Society of Agricultural Sciences*.

Wang B, Shi X, Guo C, Ye X, Wang Z and Rao P (2004) Isolation and purification of ribosome inactivating proteins from bitter melon seeds by ion exchange chromatographic columns in series.Se Pu 22(5): 543-6.

Wesley G. Taylor, James L. Elder, Peter R. Chang, and Ken W. Richards (2000), Microdetermination of Diosgenin from Fenugreek (*Trigonella foenum-graecum*) Seeds, *J. Agric. Food Chem.*, 2000, **48** (11), pp 5206–5210

Yadav, R., Kaushik, R., & Gupta, D. (2011). THE HEALTH BENEFITS OF TRIGONELLA FOENUM-GRAECUM: A REVIEW. *International Journal of Engineering Research and Applications, 1*(1), 032-035.

Yeh, G.Y.; Eisenberg, D.M.; Kaptchuk, T.J.; Phillips, R.S. Systematic review of herbs and dietary supplements for glycemic control in diabetes. *Diabetes Care* **2003**, *26*, 1277–1294.

Yibchok-Anun, S. A. (2006). Slow acting protein extract from fruit pulp of Momordica charantia with insulin secretagogue and insulinomimetic activities. *Biol. Pharm. Bull, 29*, 1126-1131.

Zhen H., X. S. (2001). *Zhong Yao Cai, 24*(2), 95-97.